EMBODIMENT AND MECHANISATION

For my family.

Embodiment and Mechanisation
Reciprocal Understandings of Body and Machine from the Renaissance to the Present

DANIEL BLACK
Monash University, Australia

Routledge
Taylor & Francis Group

LONDON AND NEW YORK

First published 2014 by Ashgate Publishing

2 Park Square, Milton Park, Abingdon, Oxon OX14 4RN
711 Third Avenue, New York, NY 10017, USA

Routledge is an imprint of the Taylor & Francis Group, an informa business

First issued in paperback 2016

British Library Cataloguing in Publication Data
A catalogue record for this book is available from the British Library

The Library of Congress has cataloged the printed edition as follows:
Black, Daniel (Daniel Ariad)
    Embodiment and mechanisation : reciprocal understandings of body and machine from the Renaissance to the present / by Daniel Black.
       pages cm
    Includes bibliographical references and index.
    ISBN 978-1-4724-1543-1 (hardback : alk. paper)
    1. Human body—Social aspects—History. 2. Human body and technology—History. 3. Human mechanics—History. I. Title.
    HM636.B526 2014
    301—dc23

2013038852

ISBN 978-1-4724-1543-1 (hbk)
ISBN 978-1-138-26724-4 (pbk)

Contents

List of Figures

Preface

This book is an attempt to capture the broad strokes of a lengthy and rich reciprocal relationship between our understanding of what a machine is, and what our own bodies are. The idea that our bodies are in some sense like machines is well established in our culture; but if this is true, what does this ultimately tell us, not only about the nature of bodies, but also the nature of machines?

It isn't hard to turn up evidence of the profound influence mechanistic thought has had on the development of modern understandings of the world. Machines are key emblems of modernity and the modern mode of solving problems: the steam engine, the motor car, the aeroplane, moon lander or personal computer are emblematic of a certain kind of society and mode of thinking and working. But the figure of the machine is also crucial to the philosophy of René Descartes, the thought of Karl Marx, and our understanding of genetics and even human thought and consciousness.

The machine – that is, not any particular machine, but an abstract conception of the machine's essence – has provided a powerful framework for understanding ourselves. It is almost certain that, had the machines of our forebears elicited different feelings or metaphors from them, the world today would be significantly different. Not only would there be differences in the kinds of machines it contained, but there would also be differences in how we understood the world and, most importantly, our own bodies and minds.

The figure of the machine has guided human understanding and development in some directions and not others. It has provided the systems of thought necessary to produce astonishing medical and technical breakthroughs, elevating its influence and confirming its value as it did so. In so doing, it has simultaneously directed us away from certain solutions to problems and certain ways of understanding ourselves, and this has also been noted and sometimes decried. Soullessly utilitarian and materialistic, treating human beings as mere objects, or oversimplifying the complexity of life and the natural world – there are plenty of grounds on which someone might take offence at the elevation of the machine to the level of a master concept to which even living human beings should be subordinated.

I am sympathetic to many of these complaints, particularly the complaint that understanding things in mechanistic terms usually means explaining a complex phenomenon according to the operation of a much less complex phenomenon (for example explaining the human mind with reference to the operation of a computer). At the same time, however, both positive and negative responses to mechanism tend to suggest that the machine has everything its own way – that the machine has become a dominating, overbearing figure that recasts everything in its own image. The problem with this tendency is that it misses the degree to which the machine is itself an unstable figure, something whose nature has not only altered inevitably over time as a result of technological changes, but also taken on attributes of the non-mechanical phenomena it has been used to describe and investigate.

There is no better example of this than the relationship between machine and body. The body–machine relationship is the most powerful example of mechanism, with an unparalleled capacity to either exhilarate us with the promise of controlling and improving ourselves or depress us with the threat of transforming our physical selves into demystified, predictable and exploitable systems. A tendency to treat the living body as a machine – a robot – is for many emblematic of contemporary society's threat to dehumanise and denigrate us. But it should also be understood that the very intimacy of this centuries-old relationship between understandings of machines and understandings of bodies has meant that, in some ways at least, the figure of the body has left its mark on the figure of the machine.

This book seeks to trace some of the ways in which both the machine has served as a metaphor for the body, and the body has served as a metaphor for the machine since the Renaissance. Our ways of understanding both have changed in important ways over time, and those changes have been crucially dependent upon the reciprocal relationship between the two. Neither figure has had a fixed significance: the very longevity of the relationship is a result of the fact that when our understanding of one term has shifted it has been able to pull our understanding of the other along with it. A belief that machines are like bodies and bodies like machines has meant that the definition of each has been importantly dependent upon analogies with the other, causing them both to shift over time in an interrelated way and without either ever serving as a reliable anchor.

The very richness and expansiveness of this relationship means that I have covered many examples of it only briefly, and omitted many more. This book is not a comprehensive history or catalogue of this relationship, but it does seek to pick out several of the more significant or interesting examples over a large expanse of time in order to demonstrate its importance and long-term significance. My hope is that, by this book's end, it will have made you more mindful of the ways in which your understanding of your embodied place in our technologised society, and your hopes and fears for how it will change in the future, are grounded in this history of reciprocal body–machine relations.

Daniel Black
Monash University, Australia
July 2013

Introduction

Avatars upon Avatars

The 2009 Hollywood blockbuster *Avatar* told a story that was fundamentally dependent upon the following belief: every living body, human or otherwise, is composed of a mechanical vehicle controlled by a computer. *Avatar* is a science fiction film, and so such a science fictional-sounding idea might seem entirely appropriate, but it is an idea that is intended to be accepted as part of the *science*, rather than the *fiction*, of the science fiction. That is, while *Avatar* is a work of imagination, it aims to make its fictional events seem plausible to an audience by building them upon a foundation of ideas that its audience already accepts to be true. The idea that living bodies are mechanical vehicles controlled by computers is part of its foundation of accepted fact, not the superstructure of speculative fiction that rests upon it. In fact, although it might be expressed using different terms in other contexts, the idea that living bodies are composed of machines piloted by computers is a widely accepted assumption in both scientific and popular understandings of the nature of life and mind.

The avatar of the film's title is a mindless alien body, an empty vehicle created in a laboratory. While it is never explained how this process has produced a fully functioning brain in the absence of the lengthy process of environmental interaction required for such a thing to exist, this avatar nevertheless somehow possesses a brain that is fully formed and functional except for the absence of some isolatable component of 'free will'. The biological robot is then controlled via some unexplained and invisible link with a computer, which serves as an intermediary between the avatar body and a human, called a 'pilot', who provides the free will the avatar lacks. The film's audience is expected to accept this fictional technology precisely because it agrees with the audience's existing beliefs concerning how bodies, brains, and computers function.

The avatar can be accepted as an empty vehicle for the human pilot precisely because real, human bodies are already accepted as being empty vehicles, utilitarian machines piloted through the world by brains encased inside them. While the body of the avatar has different dimensions and slightly different anatomy to the human pilot, the pilot can immediately take control of the avatar and be running in a matter of minutes, despite the fact that the same pilot once took over a year to figure out how to walk in his own body. The shift from controlling human body to controlling avatar body seems to be no more disruptive than the shift from driving a Ford car to a Nissan. Furthermore, this link between pilot and avatar is only possible through the mediation of computer equipment, and this is in turn only possible because both human and avatar brains are themselves kinds of computers: organic computers that can be connected to non-organic computers, transferring data to and from them without any loss or ambiguity in the signal. The brain is an information processor whose information can be sent through wires to a computer, and then through the air from computer to another brain, and then back again instantaneously and without the information ever changing its character. In order to simplify this tortuous formulation, we can cut out the intervening steps and simply say that the brain is a computer that can drive bodies; under normal circumstances it drives the body to which it is directly attached, but if it is plugged into another body it can drive that one just as well. Some of the human mercenaries in the film illustrate this idea in more direct physical terms by sitting inside giant robotic exoskeletons, which follow the movements of their bodies. These robots are like Russian matryoshka dolls, given that the giant robot body is piloted by a human body, which is, in turn, piloted by a brain riding in the cockpit of its skull.

But *Avatar* takes such ideas even further in its imagining of the alien ecology of the fictional planet Pandora. If all living bodies are ultimately controlled by computers, why not network those computers? The alien species of Pandora possess a kind of biological ethernet cable dangling from their heads, which they can use to plug into the network or make direct connections with each other. The native Na'vi can use the Pandoran horses and pterodactyls as their own avatars, plugging their brains into these creatures and turning them into biomechanical vehicles under the control of a Na'vi brain.

Outside the narrative of the film, this idea is extended further by the use of motion capture to control digital bodies. Not only does the story provide this dizzying array of examples of how bodies are machines controlled from elsewhere, but in addition the story is told using digitally animated 'puppets' controlled by the absent bodies of living actors. The now common practice of having physical and digitally rendered bodies interact in films, like our everyday experiences of seeming to project ourselves out of our bodies into the informatic environment of the Internet, creates a wider sense that machines and computers are interchangeable with bodies and brains.

In the 1980s, roboticist and futurist Hans Moravec predicted a future scenario that hinged on the same logic, in which human consciousness would be 'uploaded' into superhuman artificial bodies, which could make us immortal by freeing us from our frail flesh, which, not being an essential part of 'us', could be discarded without any impact on who we are. Moravec's enthusiastic account of how this might happen has been influential (or infamous, depending on your perspective[1]), even spurring some to begin preparations for the

---

1 For example, this account initiated N. Katherine Hayles's important book *How We Became Posthuman* (Hayles 1999: 1). In addition, the practicalities of such a procedure have been discussed in a 2012 special issue of *The International Journal of Machine Consciousness*, while its philosophical implications have been investigated by David Chalmers in the *Journal of Consciousness Studies* (Chalmers 2010), his investigation then being responded to in a 2012 special issue of the same journal.

coming transmigration. As we will see in Chapter 4, Ray Kurzweil has become the most prominent of these 'singularitarians', believing that he will one day live as a 'spiritual machine' (Kurzweil 1999). Moravec's description of the process is worth quoting at length:

> You've just been wheeled into the operating room. A robot brain surgeon is in attendance. By your side is a computer waiting to become a human equivalent, lacking only a program to run. Your skull, but not your brain, is anesthetized. You are fully conscious. The robot surgeon opens your brain case and places a hand on the brain's surface ... Instruments in the hand scan the first few millimetres of brain surface. High-resolution magnetic resonance measurements build a three-dimensional chemical map, while arrays of magnetic and electric antennas collect signals that are rapidly unravelled to reveal, moment to moment, the pulses flashing among the neurons. These measurements, added to a comprehensive understanding of human neural architecture, allow the surgeon to write a program that models the behavior of the uppermost layer of the scanned brain tissue. This program is installed in a small portion of the waiting computer and activated. The brain tissue is now impotent – it receives inputs and reacts as before but its output is ignored. Microscopic manipulators on the hand's surface excise the cells in this superfluous tissue and pass them to an aspirator, where they are drawn away ... Layer after layer the brain is simulated, then excavated. Eventually your skull is empty, and the surgeon's hand rests deep in your brainstem. Though you have not lost consciousness, or even your train of thought, your mind has been removed from the brain and transferred to a machine. In a final, disorienting step the surgeon lifts out his hand. Your suddenly abandoned body goes into spasms and dies. For a moment you experience only quiet and dark. Then, once again, you can open your eyes. Your perspective has shifted. The computer simulation has been disconnected from the cable leading to the surgeon's hand and reconnected to a shiny new body of the style, color and material of your choice. Your metamorphosis is complete. (Moravec 1988: 109–10)

The set of assumptions about the body and information that underlies this scenario is highly influential today, and follows the same logic as *Avatar*'s computer-controlled vehicle/bodies. (While Moravec's mind uploading scenario would turn its subjects into *literally* computer-controlled machine vehicles, the fact that he does not expect this to cause any change to a recipient's identity reflects a conviction that this is all we have ever been.) N. Katherine Hayles has used the term 'virtuality' to refer to a perception that our material selves are becoming increasingly absent or redundant in the age of information (Hayles 1999a: 19), and this sense of being 'virtualised' is underpinned not only by computational models of cognition but also, for example, a belief that our natures are encapsulated in stored genetic code. While our brains make us who we are, according to this account, they are still just the biological computer hardware necessary to run the software of the mind. This mind software, which exists only as immaterial information, could just as easily run on some other computer in the absence of our physical selves.

From Earthen Machines to Spiritual Machines

In 1664,[2] in his *Treatise on Man*, René Descartes invited his readers to consider how God might go about creating an artificial substitute for a human being. Given that God has been credited with creating *actual* human beings, this seems like quite an odd hypothetical situation, but presumably the religious sensibilities of the day required that Descartes begin in this way, rather than making the more presumptuous and potentially blasphemous suggestion that mere human beings could be behind such an enterprise. That the prospect of human beings simulating themselves lay behind his thought experiment nonetheless seems clear given that – in addition to the pointlessness of God's simulating human beings when he can create the real thing – Descartes imagines the simulation of human beings in terms of the human production of machines, and particularly *automata,* those mechanical entertainments that mimicked human form and action and which in later years would come to have such a hold over the popular imagination.

Introducing the business of creating an artificial human being, Descartes states:

> I assume their body to be but a statue, an earthen machine formed intentionally by God to be as much as possible like us. Thus not only does He give it externally the shapes and colors of all the parts of our bodies; He also places inside it all the pieces required to make it walk, eat, breathe, and imitate whichever of our own functions can be imagined to proceed from mere matter and to depend entirely on the arrangement of our organs.
>
> We see clocks, artificial fountains, mills, and similar machines which, though made entirely by man, lack not the power to move, of themselves, in various ways. (Descartes 1972: 2–4)

Finally, having described in great detail the similarities between human and machine that would make such an artificial body possible, he concludes:

> I desire you to consider, further, that all the functions that I have attributed to this machine ... imitate those of a real man as perfectly as possible and that they follow naturally in this machine entirely from the disposition of the organs – no more nor less than do the movements of a clock or other automaton, from the arrangement of its counterweights and wheels. Wherefore it is not necessary,

---

2   A Latin edition, *De homine*, appeared in 1662.

on their account, to conceive of any vegetative or sensitive soul or any other principle of movement and life than its blood and its spirits, agitated by the heat of the fire which burns continually in its heart and which is of no other nature than all those fires that occur in inanimate bodies. (Descartes 1972: 113)

Descartes did not believe such mechanisms to be all there was to human beings; he still held a belief in a certain kind of soul, and believed that this soul was of another order entirely. Nonetheless, this Cartesian account is the most famous example of the body as mechanical vehicle idea we see in both *Avatar* and Moravec's mind uploading scenario.

While Descartes was far from being the first thinker to see human beings as divided creatures, formed from the joining of body and soul, he is credited with popularising the influential idea that the incorporeal component of human beings is a unified essence of the self that – although Descartes still referred to it as a soul – corresponds more to a contemporary conception of the mind than the various kinds of soul or souls referred to by the philosophers of ancient Greece, for example. As reflected in his famous pronouncement *cogito ergo sum*, Descartes understood the immaterial *res cogitans* – the stuff of thought – to be the true basis of the self, while the body was nothing more than a clever machine that served as a tool for the soul/mind and was clearly divided from it.

According to Descartes' account, the body was like a car, driven around by the soul, which retained its exalted status as something beyond mere mechanics and even the ignominy of materiality. The body was no more than a collection of systems allowing physical presence in and action upon the world, which could in theory be replaced by a new machine built for the same purpose. Our real selves were of another order entirely: rarefied entities that drove this fleshy vehicle via a kind of remote control using 'animal spirits' (Porter 2003: 67).

But how could Descartes believe that his body – which was, after all, the most obvious and available part of his self – was somehow *not* him, that the crucial component of the self that made him who he was was actually hidden somewhere else? And how is it that such a strange idea became – and has remained – so influential?

Descartes' understanding of what the body is relied heavily on technological imagery. Like many of his contemporaries, Descartes subscribed to the new 'mechanical philosophy', which saw the human body as fundamentally a machine (Porter 2003: 51–2). It was a complex and cunningly engineered machine, without a doubt, but a machine nonetheless, whose lungs worked like bellows and heart worked like a water pump.

It does not follow from this that the human self can be explained entirely in mechanistic terms – far from it. Descartes' formulation rests on a belief that, owing to our possession of an immaterial soul, it is impossible to provide a purely materialistic, mechanistic explanation of who we are. However, it wasn't too long before others carried forward the movement from soul to mind, creating a wholly secular, scientific concept that nonetheless continued the theme of an immaterial component of the self that was tied to the machinery of the body (Riskin 2007: 4–5).

Descartes' claim that an immaterial soul was anchored in the material body by the pineal gland (Descartes 1988c: 230–32) struck many as implausible, and the idea that the mind was immaterial, and so of a fundamentally different nature to our material selves, separate – and potentially separable – from the body, was open to ridicule. Not long after the publication of the *Treatise on Man*, John Locke would propose another thought experiment in his *Essay on Human Understanding* of 1692:

> Let us … suppose the soul of Castor separated during his sleep from his body, to think apart. Let us suppose, too, that it chooses for its scene of thinking the body of another man, v.g. Pollux, who is sleeping without a soul. For, if Castor's soul can think, whilst Castor is asleep, what Castor is never conscious of, it is no matter what place it chooses to think in. We have here, then, the bodies of two men with only one soul between them, which we will suppose to sleep and wake by turns; and the soul still thinking in the waking man, whereof the sleeping man is never conscious, has never the least perception … For, I suppose nobody will make identity of persons to consist in the soul's being united to the very same numerical particles of matter. For if that be necessary to identity, it will be impossible, in that constant flux of the particles of our bodies, that any man should be the same person two days, or two moments, together. (Locke 1965: 82–3)

One hundred years after the *Treatise on Man*, Julien Offray de la Mettrie claimed that Descartes' division of the human subject into body/machine and soul/mind had been nothing more than 'a trick of skill, a ruse of style, to make theologians swallow a poison, hidden in the shade of an analogy which strikes everybody else and which they alone fail to notice' (La Mettrie 1927: 63–4). In other words, the establishment of a belief that the body was merely a machine inevitably set in train a line of reasoning that would ultimately render the soul – and possibly, by extension, God – obsolete.

If the human body is conceived of as a kind of machine, what need is there for a soul to guide its workings? According to La Mettrie, the soul is 'an empty word, of which no one has any idea, and which an enlightened man should use only to signify the part in us that thinks' (La Mettrie 1927: 48–9). It is an unnecessary concept once a mechanistic view of life has been adopted: the human body is no more or less than 'a machine which winds its own springs' (La Mettrie 1927: 11–12), and Descartes' soul can be cast on the scrap heap as an outdated superstition.

Mechanical Minds

These days, Cartesianism is an epithet in the study of cognition. To accuse someone of working within a Cartesian framework is to imply that their work is hopelessly compromised, wedded to an outmoded and untenable schematisation of the self. The materialist position states – like La Mettrie – that the experience of having a mind is generated entirely by physical processes inside the brain. However, in a great many cases, the person making such accusations is really no freer from the Cartesian legacy than the object of his or her criticism.

The dominant expert account of the mind, body and self takes what might be referred to as a 'post-Cartesian' position. The post-Cartesian position rejects the Cartesian account because of its materialist rejection of a Cartesian soul, but remains no more than a modification of the Cartesian view, no less dependent upon a splitting of the self between a body/machine and an immaterial and unique true self that is separable from it. However, where for Descartes this immaterial self was a divinely imbued soul, in more recent accounts it is a pattern of information.

While the machines of Descartes' day provided an explanation for how the body could pump air and blood, or generate heat or force to be exerted upon the material world, there was no parallel to be drawn between such engineering and the invisible workings of the mind, which therefore seemed to inhabit a purely spiritual plane. Rather than being simply a holdover from older beliefs or a pandering to the religious sensibilities of his time, Descartes' soul was a necessary part of his explanation. The soul filled in the gaps in his mechanistic account, providing a factor responsible for generating those faculties that he could not explain in mechanistic terms. For Descartes, it was those human capacities that could not be explained through analogy with the machines of his time – chief among them being the use of language (Descartes 1988a: 44–5) – that established the presence of a soul.

However, more recently, human ingenuity produced a kind of machine that works largely through a movement of energies imperceptible to the human eye, one whose capacity to store and process information is believed to provide a model for the workings of the human mind. As a result, from the middle of the twentieth century Descartes' mechanical body came to be joined to a computerised mind; while human beings remained split into two components in such an account, both of these components were now understood as open to technological simulation – there was no longer some excess that could not be fitted into a mechanistic account. Today the mind remains something immaterial and divorced from the body, but the advent of computing has created the idea of a similarly immaterial technology: the technology of information.

Mind Children, the book in which Moravec enthuses about the possibility of having a robot surgeon purée his brain so that he might live forever, also includes a chart comparing the computational power of such things as DNA and the Library of Congress, calculators and bees, personal computers and sperm whales, the national telephone system and the human brain, as if an ability to explain all of these things in terms of information made them fundamentally equivalent (Moravec 1988: 61). Moravec's chart originates with a belief that all phenomena that can be represented as abstract information processes must therefore be fundamentally the same, and it is a belief that the word 'information' refers to some special stuff, something imperceptible and immaterial but simultaneously objectively and powerfully real, that underlies the post-Cartesian position. Information is the modern ghost in the machine.

For Moravec, the human mind is a pattern of information, although he comes perilously close to highlighting the naïvety of this view when he remarks, 'if a mind is ultimately a mathematical abstraction, why does it require a physical form at all?' (Moravec 1988: 178). However, for him the presence of our bodies does not suggest that perhaps the mind is *not* ultimately a mathematical abstraction, but only that our bodies are outdated junk, to be jettisoned as soon as this becomes technologically feasible.

The brain becomes the house for an informatic mind that is ironically no less immaterial or disconnected from the body than Descartes' soul. As a result, the post-Cartesian position – despite its vocal disowning of Descartes – represents only a partial and compromised rejection of Descartes' position. While Descartes' soul might today seem redundant in light of materialist explanations of the brain as the engine of consciousness – as foreshadowed by La Mettrie – the fixing of our minds inside our brains has ironically become part of a larger trend towards understanding the human self as even less dependent on our physical bodies.

'Cartesianism', broadly defined, was around before Descartes and continues to the present day. The idea that the body is a machine predates Descartes – Hamlet described his body as a machine nearly half a century before Descartes did (Maisano 2007: 66)[3] – and Descartes perhaps serves more as a bridge between older, metaphysical beliefs concerning an immaterial soul and later, materialist accounts of an immaterial mind that, by doing away with the soul, are more extreme in their mechanism than Descartes ever was. While Descartes' work was a key stepping-stone on the way to a belief that 'the mind' is an epiphenomenon of the brain, for example, for Descartes the brain was still very much a part of the *machina carnis*.

Furthermore, a broader belief that we are divided creatures, produced by a marriage of convenience between a machine body and some purer, immaterial stuff is older again. It can be seen in Christianity's division of the self into exalted soul and despised flesh, and this tradition surely lays much of the groundwork

---

3 See also Sawday (1995: 146ff. and 2007: 236).

for a belief that the body is a tool for some alien motivating force. At the same time, however, it would be a mistake to see Christianity as a singular point of origin; for much of its history, Christianity's position on the body has been contradictory and open to dispute, lacking a unified position. Caroline Walker Bynum has drawn attention to the ambivalent and often contested status of the fleshy body in much Christian thought; after all, Christ's incarnation in a human body was seen as a central part of his identity and spiritual importance, and 'medieval piety did not dismiss flesh – even female flesh – as polluting. Rather, it saw flesh as fertile and vulnerable; and it saw enfleshing – the enfleshing of God and of us all – as the occasion of salvation' (Bynum 1991: 116). She elsewhere argues that 'Medieval Christianity is not dualistic in either a Gnostic, a Manichean, or a Cartesian sense (Bynum 1995: 11).

Furthermore, the clerical hostility towards the flesh mentioned by Bynum itself originates in Greek philosophical traditions that argued that, 'if we are ever to have pure knowledge of anything, we must get rid of the body and contemplate things in isolation with the soul in isolation' (Plato 1993: 128) even as the human form was exalted and glorified in Greek art and sport. (And this exaltation itself later reappeared in neo-classical Christian representation (Synnott 1992: 81–4).)

In fact, what is most striking about these themes is their continuity. While La Mettrie might have claimed that Descartes' explanation of the self was a time bomb set to destroy belief in a soul, and materialist accounts of cognition might claim to be built on a radical break with Cartesian dualism, on closer inspection these developments look more like the evolution of continuing themes than the revolutionary rejection of one explanation in favour of another.

Bodies and Machines

The new machines and technologies of their times are crucial to both Descartes' and Moravec's fantasies, but what is particularly striking is the ease with which their thinking moves from machines as analogies for ourselves – the body is like a piece of clockwork or the brain is like a computer – to the assumption that our bodies actually *are* machines, which are therefore interchangeable with other machines, and whose functioning can be understood by looking at machines rather than our bodies themselves. With the arrival of new information technologies, this belief in backward causation extends to the idea that phenomena that can be represented as abstract information processes are in fact produced by these representations, and it is in informatic and computationalist explanations of who we are that mechanism is today most prevalent.

Human beings have seemingly been driven to see themselves mirrored in successive technologies. From pumps and pulleys to clockwork to steam engines to computers, a recurring theme in the response of both those who create these artefacts and the general public has been a feeling that these technologies are evocative of the human body, and share fundamental qualities with it. But why take such a seemingly unflattering perspective on the nature of who we are, constantly belittling the incredible complexity of the human body by seeing it as interchangeable with vastly less sophisticated fabrications?

We seem to be haunted by our own appearance, seeing ourselves everywhere we look. Is this because we design new technologies to mimic us, or because we use the things we create as a point of reference when we try to explain ourselves?

These two alternative answers might suggest that either a) the fixed and inevitable truth of who we are informs how we shape the external world, or b) the fixed and inevitable truth of the technologies around us informs how we shape our view of ourselves. But the relationship is not so straightforward and mono-directional. It is clear that human beings anthropomorphise machines, projecting human characteristics onto technological artefacts in order to render their capacity for self-generated action more familiar and comprehensible. What is less clear is the degree to which we introject attributes of machines, incorporating the action of machines into our embodied experience in order to enhance our ability to interact with them and provide models for understanding the most enigmatic phenomenon of all: our own embodied selfhood. Much of what happens within our bodies is outside our conscious sensory experience, and machines have filled in much of this gap by both providing analogies for biological processes and producing new kinds of sensory data (such as medical scans, for example) that have become part of what we believe our bodies to objectively be.

Alienation from and objectification of our own bodies has existed for a long time and been expressed in a wide variety of terms. Drew Leder argues that it results from the body's innate quality of seeming 'absent' from our subjective experience (Leder 1990). But when did this objectified body come to be tied to the machine, specifically? Aram Vartanian seeks to provide an overarching history of the 'man-machine', which begins in ancient Greece (Vartanian 1973), but it would be a mistake to see the repeated use of machine analogies for life or attempts to construct mechanical bodies as motivated by a single goal or understanding of the relationship between body and machine throughout history (see Riskin 2007).

One of the key prophets of the arrival of an informatic and computationalist understanding of the human body, Norbert Wiener, recognised the historical evolution of the connection between machine and body:

> At every stage of technique since Daedalus or Hero of Alexandria, the ability of
> the artificer to produce a working simulacrum of a living organism has always
> intrigued people. This desire to produce and to study automata has always been

expressed in terms of the living technique of the age. In the days of magic, we have the bizarre and sinister concept of the Golem, that figure of clay into which the Rabbi of Prague breathed life with the blasphemy of the Ineffable Name of God. In the time of Newton, the automaton becomes the clockwork music box, with the little effigies pirouetting stiffly on top. In the nineteenth century, the automaton is a glorified heat engine, burning some combustible fuel instead of the glycogen of the human muscles. Finally, the present automaton opens doors by means of photocells, or points guns to the place at which a radar beam picks up an airplane, or computes the solution of a differential equation. (Wiener 1961: 39–40)

Just as the kinds of machines considered to be like bodies have changed over time, so the qualities considered definitional to bodies have also changed (and have in fact often arisen in response to changes in the nature of machines), meaning that the specifics of a particular historical interaction are often unlike those of other times and places. Furthermore, precisely what these posited similarities between body and machine actually *meant* – either for an understanding of machines, an understanding of bodies, or both – is often quite different. This is not an inevitable result of the objective natures of either bodies or machines, but a cultural development dependent upon them in certain ways. This process has accelerated and become more pervasive over time, through Cartesianism, functionalism, Taylorism, Fordism, computationalism, and onwards.

The design of machines and our emotional reactions to them can be seen to be fundamentally coloured by their association with bodies. Conversely, how we understand our own bodies can be seen to be fundamentally influenced by their association with technology. We see ourselves in the things we create, and the more sophisticated and seemingly internally animated they appear, the more we associate them with our own bodies. The specific technologies have changed over time – from ropes and pulleys, to clockwork, to steam engines, to digital visualisation – but the external perception of bodies has remained of crucial importance.

The following chapters will trace the broad sweep of the mutually defining relationship between bodies and machines as it evolves through time. My account is more-or-less chronological, but it has no pretensions to being a history. Because my primary aim is to discuss themes, rather than case studies, my account is unavoidably broad and generalising. In some cases, excellent historical accounts and close readings of material covered here already exist, providing invaluable information for both me and the reader interested in delving deeper; in those cases where they don't, perhaps this book will serve as an invitation for others to attempt them. What I seek to do here is look at the ways in which living bodies and machines of various kinds have come to define one another at different times, and the particular technologies that have enabled such relationships. I am not seeking to argue that the qualities of machines have come to warp our understanding of bodies, or vice versa. Neither body nor machine provides a stable term that colours our understanding of the other; instead, it now seems impossible to conceptualise one without reference to the other.

In the next chapter I will make some suggestions regarding how this strange relationship might be theorised. Then, in the chapters that follow, I will look at how new developments in how human beings have represented and investigated the human body and used technology to shape the environment they inhabit have changed our understanding of what both bodies and machines are and how we perceive them, from painting and autopsies to robots and nanotechnology. Chapter 2 will consider mechanist accounts of the body as primarily a matter of engineering, something that can be disassembled, and even reverse engineered, through the use of autopsies, diagrams, and mechanical simulations, beginning with the anatomical demonstrations of the Renaissance and moving to the automata of the Enlightenment. Chapter 3 moves from the automaton to the robot, the mechanical labourer that promises – or threatens – to replace the exertions of the human body in a nineteenth- and twentieth-century milieu in which human beings feel pressured to behave more like machines. Then, in Chapter 4, I discuss the ways in which the exaltation of information and information technologies has produced a further shift in how human body and machine are seen to relate to one another, creating a move towards the dematerialisation of both body and machine. Finally, in Chapter 5, I discuss the ways in which the promised powers of nanotechnology seem to carry this dematerialising process to such an extreme that it will ironically initiate a return of focus to the material plane – although this material plane and its relationship with immaterial information will be radically transformed in the process.

Chapter 1
How to Look at Bodies

What is a Body?

Throughout this book I suggest that our understanding of human bodies and machines are constituted in a reciprocal fashion, but what precisely does this mean? It does not mean that bodies are a kind of machine or that machines are a kind of body – although it does tend to produce such beliefs. It also doesn't simply mean that bodies and machines have things in common – which, too, is a product, rather than cause, of this phenomenon. Before you can believe that machines and bodies have things in common, it is necessary to identify certain shared attributes as definitional to both.

It is much more immediately obvious that machines have been used to define bodies than *vice versa*; the very name 'mechanism', which underlies so much of our scientific understanding of what bodies are, makes explicit the role of machines in our conceptualisation of ourselves. The more subtle ways in which the term 'machine' has come to be understood with reference to the body are more difficult to capture and less intuitively obvious, however, and in this chapter I will be largely concerned with providing a framework for considering how they operate.

In my attempts to do this, it might be assumed that I am starting from a position that could be described as broadly *social constructionist*; that is, if I believe that how we understand our bodies is determined – or at least powerfully influenced – by our understanding of artefacts created by human beings at various moments in our history, then I must believe that our sense of ourselves and our bodies is not grounded in a fixed biological reality, but rather created by cultural 'discourses'. However, this assumption would be at least partially wrong, as I do not subscribe to a belief that biology and culture form separable realms, or that any acknowledgement of the physical or biological bases of perception and understanding requires a belief in a fixed and inevitable sense of who we are.

On the contrary, our perceptions of ourselves are greatly influenced by our relationships with the other bodies and objects in our environment, not simply through culture, but also through the very physical architecture of our perceptual apparatus. Our view of other bodies is never objective, not simply because we can't gain access to anything outside culture, but at a more basic level because the physical act of perceiving is not reconcilable with objectivity. Our perception of our own and other bodies has always and inevitably been fluid and produced through its relationships with other features of our environment. The body has a significance for us that lies prior to cultural influence, and this is what gives it such power as a template against which we make sense of other things.

Importantly, whether we are considering the possibility that the way in which we view our environment is inflected by who we are, or the possibility that our environment inflects how we see ourselves, both sides of the equation are dependent upon one foundational truth: perception is never neutral or disinterested. We cannot perceive our environment in an objective or disinterested way; human vision is a utilitarian capacity that has developed to serve particular ends, and as such it is tailored to the most important uses to which it can be put. Most importantly, human vision (like all human senses, to more or less of a degree) is calibrated to the most significant feature of the human environment: the human body itself.

This fact tells us some very important things about how we perceive, certainly human bodies, but also everything else around us. There are two key points that arise from this and which will be explored at more length throughout this book.

First, it means that the human body serves as a template our minds apply to everything we see. Regardless of what landscape or living thing we survey, we are looking for the contours of a human body; the truth of this is established by the fact that, as a result, we often *see* a human body, even if it is not there. A face in the moon, the outline of a body in the limbs of a tree, human emotions or motivations in the behaviour of a dog: anthropomorphisation comes naturally to us, leading us to project human attributes onto everything around us.

Second, when we *do* look at human bodies, we see them in a way that results from the special significance of the human form. While medical and scientific scrutiny of the body seeks to produce some objective, disinterested image of the body that can be used to aid in diagnosis or explanation, the specialised form of perception for which the human sensory apparatus is tailor-made operates differently, specialised as it is for different ends. The utilisation of various kinds of machines and technological interventions in medical and scientific research itself results from a need for a literally *inhuman* mode of perception, one that lacks the specific qualities of human perception, which, while at least as perceptive and nuanced as machinic representation, nevertheless caters to conflicting aims. Attempts are made to purge the human mode of seeing bodies from medical and scientific investigation in the interests of objectivity.

The most obvious difference between the human and machine ways of seeing bodies lies in their relationship to the division between interior and exterior. While human perception is exquisitely tailored to the understanding of human bodies, this understanding has nothing to do with the kind of understanding sought by medicine and science. It is not an understanding directed towards bodily specificity or behaviour – in fact it is directed away from such considerations. While the average human being can remember and differentiate between two faces seen only briefly months before on the basis of differences of proportion measured only in millimetres, and can make accurate predictions about an individual's emotional state by synthesising multiple, seemingly insignificant pieces of data about posture, eye movement, tone of voice, etc. (Black 2011), humans have no innate appreciation for how the circulatory system works, or the dynamics of muscle contraction. In order to generate the insights and fine differentiations at which it excels, human perception must absorb the human form holistically, analysing it as a fundamentally unified entity, and must focus on the exterior of the body as it is this exterior that generates significance.

When sustained attempts were made – most importantly during the Renaissance – to map the interior of the human body through anatomy, the bodily interior appeared to investigators as a bewildering and alien landscape; for all our natural appreciation of the bodily exterior, the bodily interior was a sight unintelligible to such a mode of perception and understanding. For the vast majority of the history of the human species, the interior of the body has only been made available to human perception as a result of almost certainly fatal bodily catastrophes. The attempt to disassemble and diagrammatise the body provided the possibility of understanding – and even working upon – the human form as just one more system inhabiting our environment, but it confounded our inherited mode of looking at bodies, and encouraged attempts to nullify its influence by means of technological prostheses that could replace interested human perception with perception that was disinterested and non-human.

But the connection between our perception of human bodies and our relationship with technology is relevant to this discussion in more ways than this. I have proposed that a) how we see and understand the world around us is calibrated to the human form, and b) medical and scientific inquiry has historically cultivated a mode of looking that is heavily dependent upon technology and seeks to neutralise our inherited mode of bodily perception. These two factors can be seen to bring about a set of intertwining and reciprocal influences on the relationship between body and machine.

First, and most obvious, our tendency to understand our environment in human terms leads to an anthropomorphisation of machines. Faced with a novel artefact that requires explanation and domestication, and particularly one that seemingly has a capacity for self-movement and has been, in most cases, created for the express purpose of mimicking or replacing human labour, it is inevitable that these new technological artefacts will become strongly associated with the human body.

Second, the technologised, 'objective' gaze that medicine and science directs towards the human body produces a view of the human body itself as primarily a mechanism, a work of clever engineering composed of interlocking but differentiated systems and parts, by virtue of the fact that it is always partial and only concerned with objective physical properties. This view not only informs and even invites the reproduction of bodies through technology – making such reproduction seem possible and culturally intelligible – but it also reduces body and machine to analogues of one another, and then sets about eroding that analogy so that it seems more and more like a direct equivalence.

Third, the intertwining of these two influences produces a kind of feedback in how the human body is understood and perceived. As the body is increasingly investigated and represented in a way that depicts it as a mechanistic system, and, furthermore, other, fabricated, technological systems are seen to operate in ways that are fundamentally the same, so popular understandings of our own bodies bifurcate into a non-conscious, uninterrogated manner of making sense of bodies in everyday interactions and an intellectualised, common-sense understanding of bodies as simply mechanistic objects in the world.

Objectifying the Body

Late modern society is stuffed almost to bursting with things made to be looked at. The generation of value is strongly related to the aestheticisation of things, meaning that we live in a landscape in which human aesthetic sensibilities have been projected onto the vast majority of the things we see. The aestheticisation of these things often makes very explicit visual reference to the body (how much advertising material doesn't feature aestheticised human bodies?) and this might be explained solely in terms of practicalities. After all, if there was no body in a given advertisement, who would drink the cola or wear the jeans or drive the car?

But I think there is more to it than that. After all, in many advertisements it is the body, not the product, that is the primary focus of attention, the ostensible object of the advertisement often relegated to a minor role or – in some cases – entirely absent. Even when we consider the design of consumer goods themselves, their appearance often seems to reference the contours of the human form in more or less explicit ways. This is hardly surprising given that, for the vast majority of us, human bodies are the things we find most appealing, fascinating and desirable of all. However, having said that, it's also clear that it's difficult to draw a clear distinction between the power of bodies as objects to be looked at and the power of other – particularly fabricated – objects to fascinate. After all, the design of consumer goods relies heavily on our capacity to find inanimate objects physically attractive in a way not entirely dissimilar to human bodies, and it's not unusual to hear non-living, highly designed items like cars described as 'sexy'. Even the attraction of non-human living things, like certain kinds of – particularly infant – animals, indicates that, while the human body has the most power to attract and fascinate us, some of that power can be intentionally or unintentionally appropriated by non-human bodies and objects. We seem to have an innate propensity to see bodies wherever we look.

The body is both a subject and an object, but an important historical trend has been the evolution and influence of various ways of understanding the body as pure object, the object of a subject that resides elsewhere but has the power to animate, control and scrutinise it. This animated object is associated with other kinds of animated objects – in other words, machines. But this relationship is so close that it goes beyond metaphor – after all, metaphor requires that one term serve as a key to the other but, with machines and bodies, it is frequently difficult to identify which term occupies which position. As Evelyn Fox Keller has noted regarding computer technology:

> As computers and organisms become ever more entangled by the interweaving of ideas, skills, and vocabulary among their home disciplines and, perhaps more bewilderingly, by new modes of material construction and intervention, it becomes difficult at times to know which is serving as a metaphor for the other, or even to distinguish our descriptions of one system from those of the other. (Keller 2000: 130)

Bodies are often described as being like machines, but machines are regularly (either implicitly or explicitly) understood to be like bodies. Furthermore, and increasingly over time, descriptions of likeness have been considered unnecessary: bodies are not *like* machines, but simply *are* machines. Research into robotics, as well as the mind uploading scenario discussed in the previous chapter, are both built on a belief that at least some machines can actually *be* bodies.

The particularities of such examples and the circumstances that produced them are important areas for inquiry,[1] but I would argue that we can do more than simply plot the cultural forces that caused them to arise. Amidst such evolution and instability it is possible to state some general, unchanging principles. The most general point – perhaps too obvious for many to even take note of – is that, whatever the terms of the debates concerning the body or the role attributed to it, the body always *is* a focal point for concern and is attributed with a significance greater than itself. *How* the body is tied into an overarching system, or even what the precise boundaries and constituency of the body are believed to be, might vary, but it is never absent from human thinking. It is not possible (in fact, it's meaningless) to try to find some originary understanding of the body that predates the evolution of cultural ideas, but that does not mean that we cannot identify some biological bedrock beneath the contingencies of culture and history.

Introducing the collection *Beyond the Body Proper*, Margaret Lock and Judith Farquhar claim that, 'If bodies and lives are historically contingent, deeply informed by culture, discourse, and the political, then they cannot be summed up in any one kind of narrative. There is no clear common ground, no simple

---

1 See Hayles (1999) and Johnston (2002), for example, for an overview of the development of these and related discourses.

foundation in physical human nature' (Lock & Farquhar 2007: 2). However, I would argue that the diversity of historically contingent experience *is* all to some degree built upon a foundation of fixed human nature. That foundation does not propel us towards any one particular understanding of what the body is, but it does produce the general concern with the body noted above. The existence of this concern with bodies might seem too basic a fact to have any larger significance, but it has been a part of who we are for long enough to have influenced the evolution of the human body itself, fundamentally influencing the nature of human perception and understanding in ways beyond the exigencies of human culture and society, becoming a constituent part of the framework that shapes how such things are created. Rather than pre-determining and limiting our experience of bodies, this biological foundation is what allows the human experience of being and interacting with bodies to be probably the most varied and flexible experience we can have.

Inside Out

In both academic and popular discussions, the body is often defined by its status as a supposedly isolatable component of the self. The very existence of this concept, 'the body', is dependent upon a belief that it is a discrete object, available to the disinterested investigation not only of others but also of the separable consciousness that inhabits it. While such an isolated, instrumentalised view of the body has been criticised repeatedly in recent decades, there have been fewer voices drawing attention to the fact that this view is also dependent upon an isolating view of the mind, which is depicted as something divorced from physicality and, for all its cognitive activity, strangely passive, simply absorbing external information and only active in relation to an immaterial theatre of ideas it creates for itself.

By continuing to treat the body in isolation, attempts to redeem 'the body' often fall into the trap of accepting the traditional definition of this term; such discussions seek to draw attention to the importance of this neglected entity but, by focusing on the body, its attributes and history, they perpetuate the myth of its separateness and isolatability. In so doing, they do little to challenge the idea that the body is defined in opposition to other terms like mind, or even its external environment.

In Western traditions of thought, the body is often portrayed as fixed, recalcitrant matter distinct from the dynamism of the realm of thoughts and ideas, something whose nature cannot be altered or escaped. Today, when we think of individuals seeking to change the nature of their bodies, we're likely to think of the application of surgical or pharmaceutical technologies in an attempt to work on the body from the outside, and yet it is not especially difficult to find examples of bodies displaying an innate malleability responsive to both the environment and the behaviour of those bodies themselves.

Élite sports people, dancers, acrobats, martial artists and all manner of 'body specialists' bring about changes in how their bodies function that have nothing to do with biological destiny, but that at the same time are entirely dependent upon our biological materiality and the innate capacities and limitations of the physical self. As children we have all been both subjects and objects of the most incredible refigurings of bodily capacities: learning to walk, to talk, and making myriad other drastic changes to how we think, perceive, and physically interact with our environment. As the French philosopher Maurice Merleau-Ponty says:

> It is impossible to superimpose on man a lower layer of behaviour which one chooses to call 'natural', followed by a manufactured cultural or spiritual world. Everything is both manufactured and natural in man, as it were, in the sense that there is not a word, not a form of behaviour which does not owe something to purely biological being – and which at the same time does not elude the simplicity of animal life, and cause forms of vital behaviour to deviate from their pre-ordained direction, through a sort of *leakage* and through a genius for ambiguity which might serve to define man. (Merleau-Ponty 2002: 220)

Some of Merleau-Ponty's most striking examples of embodied experience relate to the ways in which machines – like a driver's car or a musician's organ – are incorporated into our immediate body schema and become part of our phenomenological experience (Merleau-Ponty 2002: 164–9), and such bodily experiences clearly did not exist before the invention of the car or the organ – they were called into being by new technologies and human habits.

Much of the challenge in trying to give an account of the body comes from a need to negotiate the division between body as object and body as subject. More than anyone else, the appearance of 'the body' as a focus of concern in the late twentieth century[2] is associated with the work of Michel Foucault, who, in books such as *Discipline and Punish: The Birth of the Prison* (1975) and the three volumes of *The History of Sexuality* (1978, 1985, 1986), cast the body as constituted by an institutional gaze. But Foucault's account is similar to the Cartesian tradition in its depiction of the body as the object of an alien and alienating scrutiny, while Merleau-Ponty's account presents a body that initiates action and is dynamically engaged with the world. Part of what makes the body such a peculiar thing and allows us to see it as something separate from ourselves is precisely the fact that we simultaneously *possess* a body and *are* a body; it is simultaneously who we are, and a material feature of the world that is available for our consideration and investigation. We experience embodiment, but we also work upon our bodies as material objects and possessions in a multitude of ways (for example, by dressing, styling, painting, tattooing and piercing them –

---

2 For an indication of the surge of interest in the body that began in the late twentieth century, see books like Shilling (1993) or Grosz (1994), or any number of other volumes published in the 1990s that feature the word 'body' in their titles.

which is not to suggest that there is no embodied experience associated with these things, too) and, most importantly, we interact with the bodies of others.

While there are numerous accounts of the body available, they all run the risk of coming to grief because of the same problem: it is seemingly impossible to give a unified account of what a body is or what it means. By this I mean something much more fundamental than simply an observation that the world is full of different bodies that have different experiences; I mean that one person's experience of being a body is itself an amalgam of essentially different and seemingly irreconcilable experiences. The line of thinking that underlies Descartes' account of the body is the clearest example of this problem.

Descartes' account of body and soul originates in the Cartesian philosophical method (Descartes 1988a), a position of radical doubt that arises from an experience of meditative self-reflection. He turns inwards and scrutinises his own inner mental processes, asking himself questions about their nature and how they arise. He is, in effect, thinking about the nature of thought, or trying to consciously experience the nature of experiencing. Inevitably, this gives Descartes the sense of being a disembodied intelligence, an entity of pure thought: his body seems like nothing more than raw physical extension because, during this experience, his body seems to be lying idle, inactive in the absence of any conscious direction from his mind.

> I saw that while I could pretend that I had no body and that there was no world and no place for me to be in, I could not for all that pretend that I did not exist ... From this I knew I was a substance whose whole essence or nature is solely to think, and which does not require any place, or depend on any material thing, in order to exist. Accordingly this I – that is, the soul by which I am what I am – is entirely distinct from the body, and indeed is easier to know than the body, and would not fail to be whatever it is, even if the body did not exist. (Descartes 1988a: 32)

In reality, of course, during this experience Descartes' body is anything but idle: his blood is pumping, his lungs are capturing and refining air, his digestive processes are breaking down and absorbing nutriment – even his thoughts are the result on some level of a flow of blood, chemicals and electrical signals. But it is rare for us to note these unconscious processes, and their very capacity to exist outside conscious awareness facilitates the Cartesian experience (see Leder 1990).

Descartes is an important figure in a process of objectifying the body, and this objectification of the body has been incredibly valuable and productive, most notably by spurring the development of modern medicine (see the following chapter). But the 'objective body' (Dumit 2004: 7–8) is only one dimension of our physical selves. If Descartes' account depicts the body as – to reuse the analogy I employed earlier – like a car driven around by the soul, then this self-reflexive posture is like sitting in an idling car, thinking about what it's like to drive. This, quite obviously, is an experience very different from driving itself.

If we imagine sitting in that idling car and trying to explain the process of driving, it's quite likely that our explanation would be strongly reminiscent of many influential accounts of cognition: it would highlight conscious decision-making and the mechanical control of our own bodies. It most likely would be something along the lines of, 'To turn right, I turn the steering wheel clockwise by feeding it from my left hand to my right', or, 'To shift to second gear, I ease off on the accelerator with my right foot and engage the clutch with my left, then pull the gear stick directly back with my left hand' (assuming I'm in a right-hand-drive car). The problem with this account is that it is completely alien to any experience of actually driving a car. No-one has ever learnt to drive by having someone explain to them how driving works; most people who can't drive already have this level of understanding as a result of their experience as passengers, but they can only learn to drive by actually, physically, *doing it*. If a dog runs out in front of my car, I don't make a conscious decision to stamp on the brake pedal or wrench the steering wheel counter-clockwise – the kind of decision-making suggested by an abstract account and the externalised rendering of the directional movement of objects (wheels, pedals, gear sticks, etc.) in space really has nothing to do with my experience of such an event. I see the dog and feel a stab of panic that produces a kind of convulsion in my body, and yet this convulsion is not random or independent of the context of being in a car – it somehow stiffens and contorts my body in such a way that my foot jumps from accelerator to brake and stomps down as hard as possible while my arms jerk the wheel to one side. My conscious mind is not directly operating my body as a tool any more than it is consciously selecting which keys to press on my computer keyboard as I write this sentence.

But how, then, *can* we explain the experience of driving? In terms of specifics, we can't. The reason for this is precisely that our bodily interaction with the car is not available to our conscious minds and does not result from a stream of deliberations and conscious decisions. The same is true for typing, or catching a ball that someone unexpectedly throws to us. Sitting around thinking about it afterwards, we might imagine that these actions resulted from conscious control, forcing them into the only account of action that language – as something constructed by our conscious minds – can express, but it's simply not the case.

It's important to make clear that I'm not suggesting that we operate as zombies, without thought or consciousness. Aside from moments of panic or instinct, we are still making decisions about what we do. I decide to turn my car to the left; I just don't exert any conscious control over the physical movements that bring this about. I decide to pick up a rock from the ground, but I don't exert any conscious control over the complex sequence of movements that bend my body, grasp the rock with my fingers, and raise both it and my body upwards. The obvious question to ask in the face of Descartes' account is, if the soul is a ghost in a machine, how does the ghost control that machine when it doesn't even have a detailed knowledge of how it works? If I ask myself which muscles in my arm, hand and wrist are and are not engaged in the act of picking up a cup, I am forced to concede that I have no idea at all. My only hope of finding out would be to perform the action while searching for evidence of muscular movement, perhaps by pressing on my arm at various points with my free hand. In other words, I have no greater access to the details of this process in my own body than I do for the body of a complete stranger, and must treat my body as an object.

Our experience or perception of the body shifts depending upon the circumstances and our perspective on it. Looking at someone else's body provides us with evidence quite different from that generated by our own bodies, but our own bodies provide us with different kinds of evidence depending on which actions we are engaged in and our current mode of conscious awareness. Foucault's account of the body can be seen to bring the same potential confusion: in its focus on institutional accounts of the body, it tends to substitute those accounts of massified, abstracted or regularised bodies for the characteristics of individual bodies, losing not only the experience of being a body, but even the externalised attributes of individuated, real-world bodies. The Foucauldian approach is important because it draws attention to the fact that these three levels are not independent of one another and are not necessarily even antagonistic to one another, but the ways in which they interact and inform one another are complex and need to be fleshed out by employing multiple perspectives. Without this, despite Foucault's call to investigate 'the elision of the body, anatomy, the biological, the functional' (Foucault 1978: 151), 'the body' tends to be seen from a single, unifying perspective.

It is common sense to look at our bodies as objects in the world, and try to understand them in the same terms as any other object. However, our bodies aren't just any old objects, *they're us*. Descartes' view of his body results from taking a perspective outside his physical self, scrutinising it as something separate from his true self, which is located elsewhere. As we'll see in the following chapter, he was looking at his body as if it were a cadaver laid out for scientific scrutiny, and the anatomist's cadaver had an important influence on his thinking regarding the body. But we don't experience our bodies as just objects; we experience them as the site of our interactions with the world around us. Just as the fact that my body provides the source of my perspective on the world means that it is invisible to me most of the time, so the very centrality of our bodies more generally tends to make them invisible to conscious experience; but this doesn't change the fact that they're there – that *we* are there.

Merleau-Ponty's phenomenological account of bodily experience is highly attuned to this paradoxical state of affairs. 'If my arm is resting on the table,' says Merleau-Ponty, 'I should never think of saying that it is *beside* the ash-tray in the way in which the ash-tray is beside the telephone' (Merleau-Ponty 2002: 112). We don't experience our bodies simply as lumps of matter arranged in space, which are prodded to perform various tasks by a controlling intellect.

We do not have a single, unified experience of being embodied, and much of the time we're not conscious of having any experience of being embodied at all, as our bodies are simply the ground that makes possible actions directed towards our environment. When trying to understand this, it is particularly useful to draw a distinction between *body image* and *body schema*. Merleau-Ponty took up these terms from psychological research, but philosopher Shaun Gallagher has sought to firm up the often interchangeable, unclear or contradictory manner in which they have been employed (see Gallagher 1995: 227). Gallagher has done a great deal to enrich Merleau-Ponty's account, not least by creating an engagement between it and contemporary research into cognitive function and development, and I will be here relying on his definition of body image and body schema:

> The body image is a conscious image or representation, owned, but abstract and disintegrated, and appears to be something in-itself, differentiated from its environment. In contrast, the body schema operates in a non-conscious way, is pre-personal, functions holistically, and is not something in-itself apart from its environment. (Gallagher 1986: 541)

Body image and body schema might be described as an 'outside-in' perspective and 'inside-out' perspective respectively, but it would be a mistake to take this too far. In fact, the problem of so doing draws attention to the lack of any insurmountable or unambiguous border separating the two.

> In the body image, the body is experienced as an owned body, one that belongs to the experiencing subject. In contrast, the body schema functions in a subpersonal, unowned, anonymous way .. The body image often involves an abstract, partial, or articulated representation of the body insofar as conscious awareness typically attends to only one part or area of the body at a time. The body schema, on the other hand, functions in a holistic way. (Gallagher 1995: 228–9)

In broad terms, the body image represents the body as an object of conscious consideration, as close as we come to seeing our bodies like the bodies of other people, while the body schema is how we experience the body as our interface with the world and the site of our actions. To describe oneself as tall is to employ a body image, whereas to unconsciously duck one's head while walking through a low doorway – despite being unable to see one's own head to check the necessary clearance – is to employ a body schema, but the two are not insulated from one another: our bodily experiences are one of the sources of our understanding of ourselves, and so feed into our body image, while we can only employ our body schema to drive a car, for example, after using our body image to self-consciously direct our bodies to perform the required actions until they become incorporated into the body schema and so no longer require conscious control.

Descartes' account of the body is clearly an example of seeing the body purely as image. The meditative posture that makes the body an object of conscious consideration is all about tuning out the experience of body schema and focusing on body image. This is what creates the peculiar, alienated, corpse- or machine-like quality of the Cartesian body: any sense of the body as lived is removed. As noted above, this one-sidedness is a common problem in many accounts of the body, and it is not accidental that it favours body image, rather than body schema. Body image is available to conscious consideration, it can be generalised between bodies, and it can be reduced to language in a fairly straightforward manner – in other words, it is amenable to an intellectualised account – while the body schema is not. The body schema is much more elusive: as soon as I sit down and try to figure out what my body schema feels like or how it works, I am operating on the level of body image.[3]

But the limitations of such divisions can be found even beyond the scale of our own individual bodies. Descartes' understanding of what happens when we look at *other* bodies creates a similar sense of unbridgeable division between us and them. This sense of unbridgeable division might seem inevitable – after all, different bodies are separated from one another in a much more straightforward manner than bodies and minds. But in fact we are much more corporeally implicated in the bodies we see than is often acknowledged.

The Embodied Eye

While a differentiation of body image and body schema is helpful, this is not because it introduces yet another binary opposition to how we understand embodiment. It is useful to have a finer and more accurate differentiation between different kinds of bodily knowledge, but much of their value comes from highlighting the lack of a final boundary between them; it must be remembered at all times that each informs and shapes the other. An understanding of the ways in which our senses of 'having' a body and 'being'

---

3   It is for this very reason that many important thinkers refuse to acknowledge any kind of knowledge or experience outside culture and language at all (e.g. Butler 1993: 30). However, just because we cannot discuss or consciously meditate upon a particular kind of experience without rendering it in linguistic terms does not mean that it does not have a pre- or extra-linguistic dimension. See Shusterman (2000: 120–29) for a critique of this widely accepted belief.

a body are implicated in one another is crucial to the integration of different accounts of bodily knowing, but the bringing together of internal and external perspectives on the body applies even beyond the scale of the individual. Even an opposition between our external perspective on other human bodies and our internal perspective on our own bodies can be problematised in the same way. Because this book is centrally concerned with the business of looking at bodies, an understanding of the ways in which this is not only an embodied act, but also entails an embodied experiencing of other bodies, is crucial.

The act of looking has traditionally been associated with a disembodied, isolated and meditative relationship with others and the world. However, many accounts of looking are themselves part of the tradition of thinking that obliterates bodily experience; and vision, that sense most associated with an alienated, non-interactive consumption of bodies (through the media, for example), in reality does create an embodied relationship with other bodies, even when those bodies are nothing more substantial than media representations.

Vision is not the only means by which we understand other bodies, and yet it is an important one to consider. In fact, for the last several hundred years vision has been of steadily increasing importance to how we know ourselves; whereas, for most of human history, we have seen our own bodies only as occasional, fragmentary and fleeting glimpses, today we routinely perceive our bodies as external, alienated entities in photographs, video, and so on. Furthermore, visual images of 'the body' – that is, generalised and normalised images that simultaneously refer to no particular body and every one of our bodies in particular, are firmly entrenched as a resource we use to understand and know ourselves. X-ray exposures, MRI visualisations, anatomical diagrams and so on are an important part of how we visually understand what we are on a physical level; they are, in fact, the only resources we have when forming a visual impression of the internal structures and systems of our bodies.

Vision is understood to be a particularly powerful way of knowing bodies – the use of other senses, like touch and particularly smell, have over time been demoted in the hierarchy of investigative tools available to medical science, while the ability to see has been extended through technological prostheses, not just into new scales of space and time, but also beyond the visible light spectrum and even into computer-generated images created from forms of data entirely alien to human sight (see Chapter 4). In the medical examination, employing the senses of touch and smell seems more intrusive, creating a greater sense of intimacy between patient and doctor where it is generally desired that there be a sense of clinical disengagement. Visual scrutiny – of the body but even more so of an image of that body – creates more detachment, more of an air of privacy for the patient, and neutrality and objectivity for the doctor.

In entertainment and consumer culture, the use of vision allows the mass dissemination of bodies that can be consumed in a disengaged, voyeuristic way.

The importance of the human body as image is self-evident in almost all forms of advertising and entertainment: even stripped of any other dimension of sensory engagement and existing only as plays of light on a television, movie screen or billboard, for example, the human body still exerts a power that draws attention to the importance of its visual component.

This in itself would be enough reason to draw a particular focus on the visual, but there is another, related but distinct, reason. Part of the privileged status accorded vision as a way of knowing bodies is a general sense that it is itself disembodied, that it is objective, neutral, and tethered to truth (Leder 1990: 117–19). Descartes' own work on optics provided an understanding of vision as disinterested and mechanistic, and while this account was superseded in the nineteenth century, its influence is still apparent in common-sense understanding of how vision works. According to Jonathan Crary, a model based on the operation of the *camera obscura* (forerunner to the modern camera), which informed understandings of vision from the sixteenth century, seemed 'to sunder the act of seeing from the physical body of the observer, to decorporealize vision … [T]he observer's physical and sensory experience is supplanted by the relations between a mechanical apparatus and a pre-given world of objective truth.' (Crary 1990: 39–40). A parallel can therefore be seen between Descartes' understanding of both his perception of his own body and his perception of other bodies. His mechanistic perspective not only alienates him from his own body, but from the bodies of others, as is apparent when he asks, 'if I look out of the window and see men crossing the square, as I just happen to have done, I normally say that I see the men themselves … Yet do I see any more than hats and coats which could conceal automatons?' (Descartes 1988b: 85). Descartes' mechanical eye looks out at what might be mechanical men; the eye's mechanistic reproduction of exterior surfaces cannot register internal life or soul, leaving room for doubt regarding other bodies' possession of such things.

Descartes presents the *camera obscura* as a model for how we see. Given that this book is centrally concerned with our tendency to associate bodies with machines and *vice versa*, I would be remiss if I did not stop to evaluate this particular example, although for the sake of ease and clarity I will compare the eye with the camera, rather than the less familiar *camera obscura*. Like all analogies between body and machine, the eye–camera analogy can both help and hinder understanding. Let's start with the helpful.

The most important reason why your eye is like a camera is this: your eye, like a camera, can't see. You wouldn't say, 'My eye sees a yellow flower' – you wouldn't say, 'my eye sees' anything. You see *with* your eyes; your eyes themselves don't see. Just like a camera, your eyes can register light in certain ways, and without their doing so *you* wouldn't be able to see, but they themselves don't see. Eyes and cameras can't process or understand the scene before them; they can only react to the light that illuminates a scene. After all, the photograph doesn't replace sight or render it redundant; it simply presents a new opportunity for looking.

26

But here's where understanding your eye to be like a camera can hinder understanding. Thinking about your eye as like a camera might lead you to imagine that your eye mediates between 'you' and the external environment, presenting you with a faithful, objective rendering of everything in front of it, which becomes what you then 'see'. Just as you can pull out a photograph and see there, neatly and faithfully laid out for your examination, everything that lay before the camera lens, your eyes might be understood to constantly present your brain with an objective rendering of everything in their field of vision. This is simply not the case.

Despite the claims of Cartesian optics, seeing is never neutral, objective, or disinterested. In order to function in the world, our minds assemble the various kinds of visual information produced by our brains into a unified view that we imagine to be simply whatever lies in front of our eyes. This principle is just as true for other aspects of cognition: time and again, researchers have demonstrated that different parts of our brains – particularly the left and right hemispheres – produce different kinds of information and seem to have quite different perspectives on the world. But our minds integrate these different perspectives in order to create a veneer of unity that helps us to function (unless something goes awry).[4]

Our eyes can register light, and this registering of light is, of course, objective. But this isn't seeing, any more than a camera can be said to see. The adage 'the camera doesn't lie' is true – and also utterly pointless – as a result of this fact. A camera can't lie because it simply records the scene in front of it. This also means that a photograph means nothing until it is seen by a person and – because human perception is not objective – a photograph's objective truthfulness only holds as long as no-one looks at it.

Human vision is never neutral, disinterested or objective. Furthermore, the human body is the most important key to understanding how and why this is the case. Because the human body is perhaps the most fascinating and privileged object of visual scrutiny in history, changes over time in how we have looked at it illustrate the forces that shape our vision; however, at a more basic level, I would also argue that the power of the human body is ultimately derived from innate traits that are a part of all human beings. These traits are what make the visual perception of the human body so open to varying cultural influences.

Not only is human visual perception not a disinterested consideration of an objective rendering of what lies in our field of vision, but human vision is calibrated to and organised around the privileged form of the human body. That we are constantly surrounded by representations of the human body is not simply the result of our cultural sensibilities or any conscious choice (although these things clearly play a part in the nature of those representations);

---

4   For an overview of the multiple physical mechanisms underlying our seemingly unified visual perception of the world, see Clark (2001: 25–31).

looking at human bodies and processing their appearance in certain ways is an innate part of who we are. Even when we are not presented with human bodies or representations of them, we continue to see them nonetheless because we process our visual environment using the contours of the human form as a template. Our brains are tailor-made to look at bodies and, even before we arrive at the rich influence of culture and history on how we see bodies, the very nature of human sight itself is organised around the human form.

Rather than vision being a distanced, isolating, passive or neutral mode of interaction with the environment, the act of looking, and particularly the act of looking at another body, is always active, embodied and engaged, and our own bodies are implicated in the nature and actions of the bodies we see.

I must be clear what I mean when I talk about looking at bodies. Crucially, I am not talking about the 'institutional gaze' analysed by the Foucauldian approach: this impersonal, inhuman, putatively objective gaze is produced by an attempt to remove the seer from the relationship of looking. Such a clinical approach holds out as its ideal a scientifically objective gaze, which is in no way altered or influenced by the identity of the viewer and – like all respectable science – can be repeated in an identical manner by any sufficiently trained expert. In other words, the diagnosis of pathology is not seen to be a product of the medical professional's act of looking, but rather the result of something objectively present in or on the patient's body.

This mode of viewing is clearly a highly contrived one, whose fundamentally problematic assumptions I don't need to go into here. The important point, however, is that this is not at all how we look at other people in our day-to-day lives (or at representations of people in the media, either, for that matter). Despite taking an 'outside' perspective, we ourselves, as watchers, are inextricably involved in formulating the objects of perception.

On one level, this is self-evidently true: while a doctor who has been trained for professional disinterest might experience relatively little emotional response at the sight of a horrific injury or disfigurement, for most of us our own bodies and experiences are not so easily quarantined from what we see. Bodies are – by a wide margin – the objects of perception that possess the greatest power to arouse us in every possible way, but this power results from some quite complicated internal processes that are still enigmatic, even to experts in the field.

We are not compelled to look at bodies simply out of idle curiosity or because of socialisation, and, while it might be true that our drive to do so originates in biological considerations such as mate selection or the avoidance of violent confrontation, knowing this tells us relatively little about what is clearly a far more subtle and flexible pursuit.

In fact, the division between our own bodies and the bodies we see is not very clear-cut at all. Returning to the eye–camera analogy, it is already clear that seeing is not a matter of simply mirroring the world around us inside our brains, as if our eyes were camera lenses and our brains photographic paper; already at this most basic level we are dealing with a highly selective and conceptualised view of the world, produced with reference to our external environment, rather than simply reproducing the objective world around us in our heads.

Numerous experiments have demonstrated that we don't see anywhere near as much of our visual environment as we tend to assume. This is simply because we don't actually need to see that much of the world around us in order to act in it. Rather than presenting us with a fulsome reproduction of the world, our visual apparatus simply identifies things that are important to us and allows us to recognise them in a highly conceptualised way. Vision is not a kind of window on the world, or a machine projecting a facsimile of the world onto a movie screen in our heads; this would only make sense if we believed that there was a little person inside our skulls (the philosophical term is a homunculus) sitting around watching it. In other words, such a belief only makes sense if I follow the tradition of Descartes, and believe that my self, my consciousness, is separate from my body and so needs a reproduction of the body's environment to be beamed to it wherever it may be. If I believe that I *am* my body, then I no longer need a mental model of the world because my body has direct access to the real world itself.[5]

Dealing with perception on this level requires a sensitivity to the ways in which body and environment are implicated in one another, and this is true even when dealing only with vision. Looking is never disinterested or cut off from action and engagement with the world. Looking doesn't put us in the position of peering through a mental window at things that are separate from us; the world is always partly inside us and outside us, and what is inside is determined by the posture I take to the outside world and my intentions toward and interactions with it. My view of the world is never passive, simply reflecting objective visual qualities of what's there in a neutral way.

The inner 'life of the mind' and the external environment are fundamentally implicated in one another. It would be a mistake to believe that our subjective experience and perception is a simple reflection of the world around us, but it would also be a mistake to believe that there is a clear divide between our brains and other bodies, or the rest of our own bodies. Furthermore, this blurring of internal and external, mental and environmental, is a feature of our perception of the world more generally. While accounts of cognition influenced by Descartes understand the mind as somehow divorced from materiality, even the act of looking is an embodied process.

---

5   In saying this, I am not arguing for a naïve account of perception that does not acknowledge the ways in which our bodies process stimuli rather than simply relaying objective information about our environment. However, the idea that we are 'locked in our skulls', living a kind of mental virtual reality simulation produced by our brains, is routinely overstated.

Even when we are dealing solely with the physical appearance of another body, the act of looking is never as straightforward as it seems. Actually, this is true when we look at anything, but the human body is almost certainly the most powerful example. While the *camera obscura* model of vision has long since been discarded, it is only recently that research into the neural underpinnings of vision has shown just how far from this model our visual relationship with other bodies actually is. Studies of human cognition have established that visual perception is far more complicated and fragmentary than it initially appears, and there are some key points to be made concerning what happens when we look at the things around us, and particularly how certain kinds of visual information can generate a powerful response in us. For example, the recent discovery of so-called 'mirror neurons' in the brains of humans and other primates calls into question fundamental assumptions about the relationship between vision and not only our own bodies but the bodies of others.

Mirror neurons are associated with the control of certain bodily movements, but they also respond to the sight of such movements in others, and are even activated by the sight of objects that might be manipulated in a relevant way (Rizzolatti et al. 2008). In other words, when we see another person perform an action, our own brains rehearse that action, on some level sharing the experience of the other person based on an innate understanding of their movements. It has even been shown that this system is triggered by the perception of emotions, causing us to sympathetically share in the affective states of others (Rizzolatti et al. 2008: 187).

More generally it has been demonstrated that how we perceive the space around us is determined by factors such as the range of our reach, our options for interaction with the objects in it, and its location relative to different parts of our bodies (Rizzolatti et al. 2008). Examples like these suggest that oppositions between internal experience and external world, perception and action, and even self and other, are not as clear-cut as they are commonly assumed to be. According to Mark Johnson,

> Mirror-neuron phenomena suggest that *understanding is a form of simulation.* To see another person perform an action activates some of the same sensorimotor areas, *as if* the observer were performing that action. This deep and pre-reflective level of engagement with others reveals our most profound bodily understanding of other people, and it shows our intercorporeal social connectedness. (Johnson 2007: 161–2)

Such discoveries establish the reciprocity between what we see and our own subjective experience. Seeing is never an isolated, neutral or disinterested appraisal of our environment, and this is most especially the case with our perception of other bodies. But I would like to extend this further still, first by noting that the mirror neuron system does not necessarily require another acting body – just the sight of objects one might manipulate triggers a response tied to the movements from a 'vocabulary of motor acts' that it invites (Rizzolatti et al. 2008: 46). In addition, I would suggest that our attunement to other bodies causes us to seek out – and find – the visual attributes of bodies in non-corporeal objects.

Snakes, Flowers and Superbeaks

The idea that there are basic 'form primitives', or shapes that have an innate significance for us and organise human visual perception, has been put forward already by neuroscientists. In 1999, the distinguished cognitive neuropsychologist V.S. Ramachandran published a paper entitled 'The Science of Art' with colleague William Hirstein, in which they argued that our appreciation of visual art could be accounted for by the action of various kinds of visual stimuli on the neurological machinery of sight. The response was mixed, to say the least, and debates about the neurological underpinnings of art continue today. Ramachandran and Hirstein's attempt to provide a scientific account of the human experience of art could not have failed to raise objections, particularly given their unapologetic lack of interest in art history or cultural specificity.[6] However, there is no reason why an acknowledgement of the neurological underpinnings of vision cannot be harmoniously integrated with a consideration of cultural and historical factors. After all, it is common for a continuing fascination with certain visual phenomena to be expressed in radically different ways in different places and times. The human body itself, of course, is the clearest and most enduring example of this.

To accept the idea that we are naturally inclined to respond to certain kinds of visual stimuli does not necessarily mean that certain kinds of shapes inevitably have a certain kind of meaning for us, but might simply mean that we have evolved a tendency to make certain associations with some shapes, rather than others.

Take the example of chimpanzees' fear of snakes (Ridley 2003: 194). Chimpanzees are not born with a fear of snakes; they learn to fear snakes by observing the fearful reaction of other chimpanzees. However, experiments have shown that observing the same fearful response being exhibited towards a flower will not inspire a fear of flowers in a chimpanzee; the chimpanzee learns to fear features of its environment, but it can learn to fear only features that

---

6   Strangely, the essay begins by remarking on the dismissive attitude towards Indian traditions of visual representation displayed by British colonialists; this highlighting of the fact that what is considered art and how it is responded to is heavily dependent upon the cultural background of the viewer seems out of keeping with a paper whose central aim is to identify some innate, biological basis for the experience of art.

31

exhibit certain visual qualities (Cook & Mineka 1990, Öhman & Mineka 2003). According to Öhman and Mineka,

> the fear module is differentially sensitive to different kinds of stimuli. Furthermore, although learning is an important determinant of these differences, evolutionary contingencies moderate the ease with which particular stimuli may gain control of the module. Thus, the likelihood for a given stimulus to be effective in activating the module is a joint function of evolutionary preparedness and previous aversive experiences in the situation. (Öhman & Mineka 2003: 488)

In other words, the chimpanzee has evolved a capacity to learn certain responses important to its survival, harmoniously integrating the biological and cultural dimensions of behaviour. Research on human subjects, such as that of Fredrikson (1997), suggests a similar combination of genes and experience in the development of human phobias.

One particularly striking analogy from animal behaviour used by Ramachandran and Hirstein to support their argument about art is taken from the research of the famous animal behaviourist Niko Tinbergen into herring gull chicks. These chicks, which instinctively peck at their mothers' beaks to solicit food, will similarly peck at a stick with a red stripe on it; the adult gull bears a red mark on its beak, and this mark apparently prompts the chicks' solicitations. More notably, according to Ramachandran and Hirstein's account, an unusually long stick with three red stripes on it, rather than one, elicits even more vigorous pecking from gull chicks. This leads Ramachandran and Hirstein to claim that, 'Indeed, if there were an art gallery in the world of the seagull, this "superbeak" would qualify as a great work of art – a Picasso' (Ramachandran & Hirstein 1999: 19). The suggestion is therefore that the hypothetical herring gull art lover wandering through a herring gull art gallery would find this piece of 'art' (a long brown or yellow shape with three red stripes on it) aesthetically pleasing without consciously understanding why – after all, 'it looks nothing like a beak to a human observer' (Ramachandran & Hirstein 1999: 19).

Using this example, Ramachandran and Hirstein suggest that there are certain 'form primitives' underlying human aesthetic sensibilities because of how we process visual information, and that art works by hyperstimulating our brains by playing on our sensitivity to them (Ramachandran & Hirstein 1999: 20). This is an interesting suggestion – for example, it might be that while the curve of a female breast, hip or buttock, a male buttock or pectoral muscle, or even the sweeping bonnet of a 1975 Chevrolet Corvette Stingray might all evoke differing degrees of aesthetic fascination in different viewers in different contexts, the same curved form primitive lies beneath them all.

However, even if this argument is found persuasive, there is still an unanswered question concerning how widely applicable such a principle would be. For both the gull beak and the contours of the human body, the template for the form is those other bodies that are of supreme importance to both birds and human beings; it can be seen to serve a pragmatic, innate need to fix on those other bodies that are crucial to survival. It is a substantial leap to claim that these form primitives are a function of our apparatus of vision that is independent of any *particular* objects in the world around us. Furthermore, the observation that the 'superbeak' 'looks nothing like a beak to a human observer' (Ramachandran & Hirstein 1999: 19) does not address the question of how much like a beak it looks to a herring gull.

In reality, Ramachandran and Hirstein seem to have created something of an urban myth with their account of Tinbergen's herring gull chicks (which presumably results from inaccurately recalling a reading of Tinbergen's work from some time previously) that has since been taken up more widely.[7] In actual fact, Tinbergen's models did look very much like gull heads: although two dimensional and made of cardboard, they were shaped to fit the profile of a herring gull's head and possessed a painted eye and beak. Furthermore, there was no second patch, and both the colour of the patches and their placement on the dummy head were seen to affect the response of the chicks, meaning that Tinbergen's experiment does not support Ramachandran's claims at all (see Tinbergen 1989: 29–31).[8] It is true that herring gull chicks will peck at rods or sticks with coloured marks on them (Conover & Miller 1981: 272), but they prefer to peck at more realistic representations of gull heads, contradicting Ramachandran and Hirstein's reading. While the chicks do tend to project the features of gull heads onto features of their environment, the more recognisable as a gull head something looks, the more likely the gull is to peck it.

Taking this into account, the herring gull example does not support the particular claims of Ramachandran and Hirstein, but it does still suggest that the gulls see bodies in the shapes of their environment, and it is common for humans to do the same. If any shape could be accepted as the basis of a 'form primitive' underlying the human processing of visual information, it would be the human face (see Black 2011). Numerous experiments have shown

---

7   For example, in was repeated in the BBC television documentary *How Art Made the World* (Anon. 2005).

8   However, Tinbergen did demonstrate the possibility of creating 'supernormal' stimuli (Tinbergen 1989: 44–6), which were more effective at eliciting animal behaviour than the natural phenomena that had driven their development. For example, a bird that identifies its species' eggs using the contrast created by their dark spots will choose to incubate a dummy egg with darker spots in preference to a real egg, and a bird that responds to egg size will choose the larger egg of another species in preference to its own if presented with a choice (Tinbergen 1989: 44–5).

that infants have a fascination with faces that does not come about through education or enculturation, and that seemingly serves a biological function in the development of social and communicative skills (just like the infant's innate predisposition to absorb language). Infants have been shown to recognise faces within days of birth (Slaughter et al. 2002: B71), even when they are presented with only a rudimentary line drawing. Very young infants show a preference for accurate representations of the face over those in which features have been rearranged randomly, even though such a preference is not displayed towards representations of the body as a whole until the age of eighteen months (Slaughter et al. 2002: B78).

Consider the iconic 'smiley face' image. A circle with two dots in its upper hemisphere and a curved line in the lower, it's a rudimentary rendering of the complexities of the human face, to say the least, and yet we can effortlessly identify this shape as a face and may very well experience some aesthetic response to it; all this despite the fact that a herring gull cognitive neuropsychologist would be likely to complain that this image 'looks nothing like a face to a herring gull observer'. In other words, if certain forms are hardwired into our brains as having a special significance (and there is good reason to believe that this is the case, given the gull chick's response to beak markings and the human infant's response to faces), then whether superbeaks or smiley faces look objectively like beaks or faces is hardly relevant. As noted above, seeing never *is* 'objective': we see a face in a rudimentary collection of lines because our brains naturally seek out any visual information that can be assembled into a face; it's not unreasonable to assume that herring gulls perceive a beak in the form and colour of the 'superbeak' shape, too. If the superbeak is understood to be like a gull Picasso, it should be for this reason: if we look at one of Picasso's cubist renderings of the human face, what we see looks vastly dissimilar to what 'objective' photographic representations have led us to believe a face should look like, and yet we have no difficulty recognising it as a face.

The superbeak analogy does not, therefore, establish that these 'form primitives' exist beyond the reach of conscious awareness as visual stimuli we enjoy without knowing why we enjoy them; in fact, our very obsession with certain kinds of visual stimuli makes us prone to identify anything that chimes with that form primitive as a face, for example, rather than being unaware of the face's significance as an object of perception. Even with the contours of the body more generally, we appreciate these shapes in actual bodies or representations of them (whether in art, real life or elsewhere), and where they appear in more abstract forms their significance remains close to the surface. After all, it's no accident that a 1975 Stingray is described as a 'muscle car'.

Research into how our brains function like that discussed above can provide evidence that we both internalise what we see and project outwards onto our environment certain naturally privileged shapes. At the same time, however, most research into how our brains function is informed by the same assumption that underlies the mind uploading scenario, and which I will later critique in more detail – that is, that we are reducible to patterns of information generated by computational processes in our brains. In Chapter 4 I will deal further with this assumption, but will here note that much brain research is therefore of value not just because of what it might say, but because of how it is conducted. In the past several decades, new scanning and imaging technologies such as fMRI, CAT and PET have facilitated huge leaps in our understanding of how our brains work, and without these new technologies the current discussion would not be informed by useful scientific insights into how we process visual information. However, at the same time, these technologies *themselves* produce visual information, which itself becomes a part of how we understand and perceive our bodies and the bodies of others. An fMRI scan itself produces a certain way of seeing and a particular perspective on our bodies, one that was not previously available to the human eye. Its significance therefore not only results from the information it might give us concerning how we look at bodies, but is also a product of the fact that it is *itself* a part of how we look at bodies. Like still and moving photography, medical dissections, portraiture and phrenology before it, it is a technology designed to investigate our bodies by producing visual information about them. The historical relationship between new technologies – primarily but not exclusively technologies of seeing – and our understanding of bodies is therefore of great importance to the account of how we see bodies that follows.

Looking at Machines and Seeing Bodies

While we might believe that our perception of a sports car or new pair of shoes can safely be discussed primarily with reference to the realm of culture alone, the human body clearly straddles the nature–culture divide in a way that few – probably no – other objects of perception can. The reason for this is obvious: while human bodies are of central concern to culture – being investigated by medicine and art, decorated and modified by fashion, and disciplined and organised by codes of etiquette and criminal law, to give only a few examples – our perception of and interaction with bodies is also of crucial biological significance. Our ability to understand and respond effectively to the behaviour and appearance of other bodies is of importance at a very fundamental, biological level; an inability to do so can be expected to seriously endanger any human being from the moment of birth, and this consideration can be seen to underlie the easily verifiable fact that human brains have evolved to respond to human bodies – and especially faces – in certain predetermined ways. Clearly, the human body has a privileged position as a powerful locus of significance when looked at from both the cultural and biological angles;

this inevitably leads to the question of precisely how the two interact. To what degree is our perception of bodies an innate biological given, and to what degree is it a product of culture? The powerful influence of both sides makes this a particularly knotty question, and one for which no final answer seems available. All that can be done, given our current understanding of both cultural history and brain function, is to identify and discuss some of the factors at work on each side of the equation in the hope of getting a general sense of how they influence one another.

The representation of the body, whether through painting, sculpture, medical case study, literary description or computer imaging, is another layer of processing that overlays what takes place within our own cognitive apparatus. Neither level has access to some objective reality of the body (or any other part of the physical world), but cultural representations and cultural understandings begin with the already-processed material produced by our innate faculties. Clearly, there is an interplay between the two: how we represent or understand bodies within a particular cultural or social context must be strongly influenced by the information our apparatus of perception provides us with concerning those bodies, and it seems reasonable to assume that – to a degree and in a way that is harder to specify – our cultural and social context will have an impact on the development of the perceptual systems we use to understand the bodies of others.

And this does not just apply to the bodies of others, of course. As noted above, we are all able to scrutinise our own bodies as if they were something separate from ourselves. Jaques Lacan's image of the child's becoming a social subject by looking into a mirror and seeing his or her own body as it is perceived by others is probably the most powerful depiction of this relationship (Lacan 1977: 1ff.).[9] Those who blame thin fashion models for eating disorders among young women might be oversimplifying the factors that produce such problems, but the very existence of this accusation reflects a belief in the power of representations to influence our lived experience of our own bodies.

While the human form is an important – perhaps the single most important – feature of the visual landscape in which we live, in many cases this form comes to us as a representation rather than a living body. It seems unlikely that our perceptual apparatus has evolved to make strong differentiations between representation and living body, and in fact if this were the case then presumably artistic representations of the body would enjoy far less significance for us than they actually do. Furthermore, very young infants seem to recognise drawings

---

9   That said, it should also be noted that the idea of infants initially having no sense of themselves as integrated corporeal entities, which was influentially put forward by mid-twentieth century research in developmental psychology and which Lacan's mirror stage relies upon, is now open to question in light of more recent research (see Gallagher 2005: 83–4).

of faces, not just living faces (H.R. Wilson et al. 2002: 2910), suggesting that whatever innate ability to identify faces they are born with does not differentiate between the two. This of course does not mean that we respond to both living bodies and representations in exactly the same way – after all, no representation will be able to produce all the kinds of stimuli we gather from someone's body during real, face-to-face interaction – but it reflects the fact that even representations have a particular power to stimulate us, and I can certainly think of no evolutionary reason why our brains would have developed systems to prevent this from happening.

Our perspective on other bodies is so fundamental that it colours our perception of far more than actual bodies themselves, lying at the heart of how we make sense of what we see. This itself establishes that our brains' systems for dealing with what we see are calibrated to the specific task of reading information from the surface of the human body.

Human beings are most importantly social creatures, and interaction with other bodies is a central aspect of human experience. Even the most cursory survey of the mass media or fine arts is enough to convince us (if we need convincing) that we have a deep personal investment in the appearance of bodies, and a craving to draw the experience of other bodies into our own selves. Human beings, as social animals, have evolved a capacity to appreciate and absorb the attributes of human bodies so sophisticated that we are only not astonished by it because it is such a fundamental, unquestioned, and constant component of our lives.

My larger conclusion is therefore that we look for bodies everywhere, and have an innate propensity to perceive bodily attributes in our environment. Our own sense of being bodies extends out into the world, which we understand to an important degree in bodily terms, and we constantly search for corporeal meaning. But why are machines of particular relevance to all of this?

This will of course be addressed in the following chapters, but I will here indicate two crucially important attributes of machines. First, machines are animated – that is, they move themselves, and when technology reaches the required level of sophistication, generate their own energy, something which, at least since Aristotle, has been associated with life. Second, machines are as a general rule designed to reproduce, replace or complement the movement of living bodies. The usefulness of machines arises from their capacity to either magnify the efficacy of bodily movement, or take its place; their function and design has historically, therefore, been given meaning by bodies. Both of these qualities suggest a relationship between the animated object and the objectified body.

Machines, which give the impression of internal animation, call to mind living things, and, given the fundamental role of the human form in our perception of the world, it should hardly be surprising that novel creations are filtered through the lens of our anthropophiliac eyes. However, the interplay between our understandings of machines and the human body can be seen to extend much further than simply desiring machines that simulate our own form and movement.

We consistently anthropomorphise machines, our attempts to conceptualise unfamiliar new artefacts falling back on the most fundamental and sophisticated frameworks for understanding animation we have – those related to the human body – and our attempts to reconcile ourselves with the unfamiliar or uncanny lead us to imagine these interlopers as the most familiar and recognisable forms we know – ourselves.

But that isn't the end of it – in Western history, at least. The act of recognising our own bodies in the functioning of machines has resulted in a propensity to see machines as metaphors for our own bodies. Looking at machines and finding them evocative of our own bodies, there has been a tendency to conclude that this results from some fundamental property of the machines themselves, rather than being a product of the fact that our brains are designed to see the human form in our environment. This has often resulted in a paradoxical state of affairs where machines are understood to provide insight into how our bodies work, and held up as a kind of template for the human body. The result has sometimes been a feedback loop where body and machine have had a reciprocal effect on our understanding of each other, our conceptualisation of new machines heavily influenced by our existing understanding of our own bodies, but our understanding of our own bodies then, in turn, being inflected by our understanding of these machines. This is already apparent in the mechanical philosophy of Descartes: his mechanistic account of the human body relies on metaphorical associations between the lungs and bellows, or the muscles and rope-and-pulley assemblies, for example, but does not acknowledge the fact that bellows were created to mimic and replace the action of the lungs, or that rope-and-pulley assemblies were created to assist the operation of human muscles. The parallels between the human body and these mechanisms is neither accidental nor the result of some natural law, but rather arises through an intentional reshaping of the environment to complement the human body. The most powerful example of this tendency is the process by which the computer, a device created to mimic a certain human mental capacity (the capacity for computation), comes to be treated as a dominant metaphor for the operation of the brain and thought more generally.

The tension between the body and embodiment, the body as object and subject, sustains the paradoxical nature of this relationship. The body as an external form is the most familiar, comprehensible and appealing thing in our world, and thus functions as a foundation for our understanding of machines; but at the same time, the interiority of our bodies and the systems that govern their living processes are enigmatic and confusing to us, leading us to employ the familiar principles behind the operation of machines as well as mechanised forms of measurement and representation in our efforts to understand them. Because the machines we build do not actually contain some truth about our own bodies – they only give this impression because our minds look for bodies everywhere around us – there is always a slippage in the movement from one term to the other. The machine being described as like a body is never *quite* like a body, and the body being described as like a machine is never *quite* like a machine. As a result, with each reciprocal movement properties not appropriate to one or other of these terms are carried over into our conceptualisation of them: properties of machines become attached to the body, and properties of the body become attached to machines. Ironically, this in turn magnifies our perception of similarity or even interchangeability between the two, encouraging further exchanges.

To suggest that our relationships with machines are built on innate human propensities does not at all mean that such relationships are inevitable – it doesn't even mean that the same kind of relationship could not arise with some entirely different feature of our environment. The relationship between body and machine is dependent upon the history of certain ways of thinking about bodies and certain ways of thinking about machines, and this history draws together numerous different ideas and influences. Furthermore, this history is dependent upon the kinds of machines being made at particular times and the roles they have been attributed with in society. As a result, the particularities of this relationship have inevitably changed through history, and the following chapters will seek to create a sense of how particular ways of thinking about the body and particular kinds of machines together have inflected this relationship over time.

Chapter 2
Machina Carnis

Making the Body an Object

No bodily experience is entirely internalised and isolated from external factors, nor is it entirely external and isolated from individual specificity. We understand the bodies that we see through our own embodied human perception. We understand both our own bodies and the bodies of others through the interaction of bodies, or the interaction of one body with itself (since it is possible to experience one's own body as a feature of the environment as well as the locus of experience).

The action of mirror neurons (see previous chapter) demonstrates that our own bodies, minds, and capacity for action respond in an embodied way even to the perceived actions (and potential for actions) of other bodies. But if we accept this characterisation of the human body as an object of visual perception, where does this leave the idea of objectively knowing the body? The objective, neutral, disinterested investigation of the human body has been a key project of modern society, whether it be in anatomy, medicine, time-and-motion studies aimed at increasing industrial efficiency, or genetics. However, if the very act of looking at another body engages a pre-conscious apparatus whose role is to process and conceptualise that visual information, and our own bodies are always implicated in the actions of other bodies, how can it even be possible to adopt a disinterested, objective position in relation to another body?

Perhaps it isn't possible at all. However, for the moment I'm more interested in how it might be possible to go about *trying* to achieve this than whether or not such an attempt can ultimately succeed. After all, for at least the last half millennium, human beings have been engaged in an attempt to objectively know the body, and this project has required the development of all manner of techniques and working assumptions, and produced all manner of benefits, regardless of whether or not it is founded on an impossible desire.

Any attempt to see the body objectively requires that the pre-conscious, interactive and reciprocal perception of bodies that comes to us naturally be short-circuited in some way in an effort to remove the hidden layer of perception that implicates the subjectivity of the viewer in the body being looked at. An obvious way to bring about this short-circuiting would be to employ a defamiliarising and fragmenting view of the body. One could attempt to reduce the body to parts in order to frustrate our perceptual apparatus's attempts to integrate the various features of what one sees into a single unit. Even better would be an approach that provided views of the body or parts thereof that were alien to those views to which human beings have been traditionally exposed. In other words, if one were to actively seek to avoid as much of this pre-conscious processing as possible, one would take an approach to the body exactly like that which *has* traditionally been taken by modern Western science and medicine.

The Rebirth of the Body

The second essay of Nietzsche's *On the Genealogy of Morals* seeks to create a kind of origin myth for modern Western civilisation. Fundamental to its concerns is the question of how social forces *get into* the body as a physical, biological entity. Indeed, Nietzsche opens the essay as follows: 'To breed an animal *with the right to make promises* – is not this the paradoxical task that nature has set itself in the case of man? is it not the real problem regarding man?' (Nietzsche 2000: 493). Not really how anyone would frame the problem today, perhaps, but nevertheless it captures the problem of how an animal can arise that is as comprehensively integrated into social structures as a human being. Nietzsche understood that culture does not exist on a separate plane from the body, and certainly doesn't render the body meaningless or irrelevant as social constructionist accounts might suggest, but neither is it the point of origin for ways of thinking and behaving that arise from it. Rather, cultural factors must be anchored in the body somehow to produce an integrated set of sensibilities; phenomena that run counter to those sensibilities must inspire revulsion or loathing in our bodies in the same way that phenomena we have an innate aversion to – the smell of excrement, for example – do. And Nietzsche suggests that this can be a grisly business, precisely because it entails working upon the physicality of human flesh itself.

For Nietzsche, the admirable product of this process was the sovereign individual (Nietzsche 2000: 495), the man of his word who took responsibility for his actions and obligations, but he suggests that this man was the product of a history of cruelties and punishments that 'domesticated' human beings so that they became able to control themselves in the absence of any direct threat of physical reprisal. There is no need for me to go into his argument in great detail, or even to evaluate its persuasiveness as a historical account; what is significant is that he argues, in the broadest sense, that society as we know it is dependent upon the cultivation of sensibilities in human beings, and that cultivation begins with that locus of human experience that predates culture and history: the human body as a biological organism capable of non-rational sensations such as pain (and other sensory experiences that Nietzsche here seems less interested in).

This essay has proven very influential. The idea that the body becomes socially productive through a process of taming and training underlies Foucault's work on the body (Foucault 1977), and is also picked up by Deleuze and Guattari in their account of how the human body becomes 'civilised' (Deleuze & Guattari 1983: 139ff.). But Deleuze and Guattari put a less positive spin on civilisation than Nietszche, and their writings are largely concerned with the possibility of escaping its effects. Their model of a body that escapes society's efforts to hollow it out and create an interiorised subjectivity is 'the body without organs' (Deleuze & Guattari 1983: 9ff., Deleuze & Guattari 1987: 149ff.). It suggests an experience of the body as a pure, open externality generating sensation through its constant interplay with its environment. This might seem like a fanciful or abstract description of any body, but on reflection it is much closer to how we experience our own bodies and those of others than the descriptions you find in a medical textbook. If I look at or touch another human being, I do indeed experience that body purely as an exteriority. If there is an internalised subjectivity within that body, it is beyond my direct experience. At the same time, all my sensations and experiences occur at the level of my own exterior: I might accept the truth that, while I am writing this, my stomach is digesting a meal and oxygen is passing through the walls of my lungs, but I have no direct, conscious experience of these things. I know my heart is beating and my lungs working, but I know this only because of the passage of air over the surfaces of my throat, the rise and fall of my chest, and the pulse reverberating through my body. Our natural, experiential knowledge of bodies is not tied to the internal details of our anatomy. If this were not the case, medical technologies such as X-ray imaging would be unnecessary, as we would be able to sense directly what was happening inside us. To claim that the representation of a body found in an anatomical treatise is more accurate than one formulated by treating the body as pure exteriority is like arguing that a diagram of the molecular structure of wood is a more accurate representation than a photograph of a tree: they are both equally accurate as representations in their own ways, although the latter has the advantage of tallying with our own perception and experience, while the former does not.

If there is a 'modern' perspective on the body, it seems characterised by a tendency to break it down into fragmented constituent parts, or to capture it at inhuman spatial or temporal scales. The 'civilised' body discussed by Nietszche, Deleuze and Guattari, or Alphonso Lingis (1983), appears unified and organised through an internal structuring, no longer requiring the direct application of torture or tribal body modification, for example, to make it part of social structures, and yet today the body is worked upon in all manner of ways that, even if different in nature from the bloody interventions of ancient times, seem no less extreme. The snip and slice of scissors and scalpels, the squirt of Botox, the Viagra-induced Niagaras of blood, the gush of saline into breast implants, the clack of ceramic and stainless steel joint replacements, the rush of air from respirator into lung, the slithering of various probes into myriad cavities and lumens, the blast of lasers and invisible beams of radiation into tissue – our society can hardly be said to leave the body to its own devices. The body is worked upon more extensively than at any other time or place, but both the different nature of this work and the very fact of its extent reflects the historical development of an idiosyncratic way of seeing the body.

The Renaissance is a historical period likely to loom large in any discussion of the birth of modern medicine and science, the depiction of human bodies, or the history of representation and ways of looking. One doesn't have to mythologise it as a time of 'rebirth' isolated from the historical forces from which it arose to see it as a time of great change and foment that has had a profound affect on human history. Michel Foucault has established that a key component in the creation of the modern Western citizen was a change in the relationship between the human body and society, and although his work focuses upon the Enlightenment, work that predates Foucault draws attention to the appearance of these themes during the Renaissance. In fact, a number of different but interrelated themes arising from the changes of the Renaissance converge on the human body and its shifting significance.[1]

The beginnings of modern science and art during the Renaissance changed how the body was represented and understood, but the powerful social changes of the time were also transforming how the body was lived in an interrelated way. In his seminal work *The Civilizing Process*, Norbert Elias claims that, during the Renaissance,

> The increased tendency of people to observe themselves and others is one sign of how the whole question of behavior is now taking on a different character: people mold themselves and others more deliberately than in the Middle Ages.
>
> Then they were told, do this and not that; but by and large a great deal was let pass. For centuries roughly the same rules, elementary by our standards, were repeated, obviously without producing firmly established habits. This now changes. The coercion exerted by people on one another increases, the demand for 'good behavior' is raised more emphatically. All problems concerned with behavior take on new importance. (Elias 2000: 79)

Elias argues that a key factor in the creation of the modern, 'civilised' individual was the cultivation of good manners – the dissemination and adoption throughout society of certain standards of deportment and sensibilities

---

1   Of course, a modern Western conceptualisation of the body is not the product of any one historical period; it results from slow, diffuse processes and varied influences (see Mellor & Shilling 1997: 42; Stallybrass & White 1986: 85). I have here chosen to focus upon the Renaissance as that historical moment that best marks the appearance of changes central to this book.

regarding the body – and the above quote identifies themes of scrutiny (and increasingly *self*-scrutiny) and the inculcation of habits as key to this process.

For the up-and-coming urban classes of the Renaissance, the body became a form of social capital; with a loosening of feudal structures focused on the inheritance of position, newly acquired wealth allowed some people access to a more rarefied stratum of society, but in order to function there they needed to cultivate the proper behaviour and refined sensibilities of their new social circle (Synnott 1992: 90). Etiquette, which had once only been the concern of courtiers, was disseminated through a wider social sphere of people who wished to behave in a courtly fashion. Manuals of behaviour became popular as individuals set about trying to refashion their sensibilities to reflect the qualities associated with 'good breeding'.

What this suggests is that bodies – the bodies of others but importantly also individuals' own bodies – were increasingly objectified at this time. Of course, the fifteenth-century man about town still had embodied experiences, but he was increasingly aware of how his body looked to others. This in turn led him to look at his body as he imagined others saw it, and to modify its appearance and behaviour when he deemed this appropriate. Where social status was once a fixed thing inherited from one's parents, it could now – for a lucky few – become a matter of show, and thus could be modified to the degree that the body as an object of perception could also be modified.

Returning to Gallagher's account of the relationship between body image and body schema discussed in Chapter 1, we can see that a consciously or unconsciously directed shift in body image can have a lasting impact on body schema. Parents instruct a child to consciously observe certain expectations of deportment and action (to close one's mouth while eating, or to not play with 'dirty' things) but, in time, the child behaves this way without conscious self-direction, and is even repulsed by the 'dirty' things he or she had previously found fascinating. During the Renaissance, a substantial number of people directed such a process of modification towards themselves, consulting reference works on those bodily habits associated with the upper echelons of society and self-consciously incorporating them into their personal behaviour. The fact that the etiquette tips discussed by Elias frequently make for amusing reading today highlights the degree to which we have internalised sensibilities that were once novel to the general public.

> [I]t is not a refined habit, when coming across something disgusting in the street, as sometimes happens, to turn at once to one's companion and point it out to him.
>
> It is far less proper to hold out the stinking thing for the other to smell, as some are wont, who even urge the other to do so, lifting the foul-smelling thing to his nostrils and saying, 'I should like to know how much that stinks', when it would be better to say, 'Because it stinks do not smell it'. (Elias 2000: 111)

According to Daniela Bohde, this change in sensibilities was also marked by an increased concern with the skin as the boundary of the self in relation to the spread of disease and other interpersonal anxieties (Bohde 2003: 31–2).

The Soviet literary scholar Mikhail Bakhtin's work on the Renaissance writer Rabelais similarly argues for an important shift in the status of the body at this time, a change in ideas about embodiment intertwined with and reflected in artistic representation. For Bakhtin, 'grotesque' bodily figures such as Rabelais' Gargantua reflect a mediaeval celebration of bodily excess:

> [T]he artistic logic of the grotesque image ignores the closed, smooth, and impenetrable surface of the body and retains only its excrescences (sprouts, buds) and orifices, only that which leads beyond the body's limited space or into the body's depths .. Actually, if we consider the grotesque image in its extreme aspect, it never presents an individual body; the image consists of orifices and convexities that present another, newly conceived body. It is a point of transition in a life eternally renewed, the inexhaustible vessel of death and conception. (Bakhtin 1984: 317–18)

In a complementary historical account, Philip A. Mellor and Chris Shilling have sought to characterise a mediaeval body largely understood through its relationship with mediaeval Catholicism. According to their account of the mediaeval body, while small élite communities (such as monks, holy women and hermits) did seek to radically control their bodies at this time, the wider community's fleshy predispositions were more likely to be assimilated into religious ritual by a church lacking the power to deal with them more strictly. Rituals of eating and drinking (of which Communion must be considered exemplary), and more extreme bodily religious rituals, such as the carnival, baptism, flagellation, starvation and other forms of bodily mortification, can be seen as a process of negotiation with the body (Mellor & Shilling 1997: 37).

According to this account, then, ritual events in which the body played a central role, such as Communion and carnival, were sanctioned by authorities because members of the general populace could not yet be expected to ignore or discipline their own bodies in the manner of the modern subject. As a result, the unruly bodies of the general citizenry were incorporated into ritualised events that channelled their appetites and desires in productive ways.

> The Church ... addressed the bodily immersion in the natural, social and supernatural worlds of the medieval era by seeking to *harness* these somatic experiences to the development of its own, sacred, communities ... Instead of trying to eradicate the manifestations of grotesque bodies that appeared resistant to control and hierarchy, ecclesiastical authorities directed them towards the effervescent experience of religious issues through the Church's organisation of carnival. (Mellor & Shilling 1997: 64–5)

The grotesque celebration of the body discussed by Bakhtin can therefore be understood as part of a preliminary effort to organise and control the bodies of the mediaeval citizenry. With members of the general population not cultivating and disciplining their own bodies, those bodies had to be addressed by and involved in rituals that created community. However, during the course of the Renaissance, the 'polite', élite orders of society became more widely concerned with cultivating and disciplining their own bodies, evincing a distaste for the 'dirty' or 'grotesque' body and its processes. Soon only the 'lower orders' continued to revel in bodily grotesquery, and such concerns were marginalised and frowned upon.

At the same social moment that Elias's *nouveau riche* merchants are poring over their manuals of etiquette and changing their bodily habits, the grotesque sensibility is joined and ultimately supplanted by a view of the body that, while fundamentally modern, is charged with a connection to the classical culture whose 'rediscovery' was credited with bringing the Renaissance about. The riotous and impolite grotesque body was elbowed aside by the 'classical' body; a body that was unified and individuated, clean, smooth and beautiful.

> The new bodily canon presents an entirely finished, completed, strictly limited body, which is shown from the outside as something individual. That which protrudes, bulges, sprouts, or branches off (when a body transgresses its limits and a new one begins) is eliminated, hidden, or moderated. All orifices of the body are closed. The basis of the image is the individual, strictly limited mass, the impenetrable façade. The opaque surface and the body's 'valleys' acquire an essential meaning as the border of a closed individuality that does not merge with other bodies and with the world. All attributes of the unfinished world are carefully removed, as well as all the signs of its inner life. (Bakhtin 1984: 320)

This new, 'classical', body comes to usurp the unruly grotesque body, and it is, of course, an idealised version of the very body Elias sees inventing itself in the everyday life of polite urban society.

Bakhtin himself explicitly links the artistic representations of the body with which he deals to the social changes of the time:

> Similar classical concepts of the body form the basis of the new canon of behavior. Good education demands: not to place the elbows on the table, to walk without protruding the shoulder blades or swinging the hips, to hold in the abdomen, to eat without loud chewing, not to snort and pant, to keep the mouth shut, etc.; in other words, to close up and limit the body's confines and to smooth the bulges. It is interesting to trace the struggle of the grotesque and classical concept in the history of dress and fashion. Even more interesting is this struggle in the history of dance. (Bakhtin 1984)

This new body is most apparent in the neo-classical artistic representations of the Renaissance. Michelangelo's *David* is probably the most iconic example of this ideal, far removed from representations of the body as permeable, disordered and open to the world. Although classical Greek thought contained several different – sometimes opposing – strands of thought concerning the status of the body, neo-classical art 'rediscovered' the body as it was idealised and glorified in the classical world, as the individuated property of the free citizen, a possession to be cultivated, cared for, and admired. In the words of Anthony Synnott,

> The Renaissance … re-discovered the body, and transformed attitudes to the body. Artists like Botticelli, Leonardo da Vinci, Michelangelo, Raphael and Titian painted the body as beautiful and in glowing colours. Cellini gave new poise to sculpture. Philosophers like Castiglione praised beauty as a 'sacred thing' and 'a true sign of inner goodness'; 'the good and the beautiful are identical, especially in the human body'. (Synnott 1992: 90)

However, where artistic representations of the body had previously dealt with characters from religion or mythology, now they increasingly also captured the individuality of real, living people, reflecting a change in the status accorded to the modern citizen.

> The individual was no longer a member of a community in the sense that medieval man had understood this. He became a body all to himself. Individualism, where it appeared, introduced the image of man enclosed in his body, his mark of difference, especially through the epiphany of the face. (Le Breton & Walker 1988: 54–5)

A fundamental change in how the body was experienced and imagined was taking place, then – as striking as the difference between a Bosch and a Botticelli. The body was decreasingly understood as part of a larger agglomeration of humanity, interconnected with other bodies and incorporating and emitting various forms of matter, and increasingly understood as separate, self-contained and self-determining. It was a valuable possession of the individual, to be cultivated, cared for and improved; the matter of what went into the body and what came out, and the means by which such potentially distasteful transactions occurred, was increasingly hidden or circumscribed by etiquette. Returning to the question of whether the body is experienced as an exterior of sensation and observation or an interiorised individual, it might be concluded that, at this moment, the emphasis was upon the exterior; after all, the body's status as possession and object of observation is highlighted, while the body's permeability is repudiated.

However, this is not really the case. First and most important, this body is understood to be only the surface or possession of an individual agent who resides somewhere 'inside' it. This is an important moment in the birth of the modern citizen, an individual who is self-determining and responsible for his or her own actions, the bearer of a unique inner life. In its focus on the interiorised individual and this individual's interior experience, it is a view very much concerned with depth, rather than surface (see Trilling 1972). A cultivated exterior, or the wearing of fine clothes, are external expressions of an interior personality, and artistic representations of the body seek to make its interior truth manifest upon its surface. Furthermore, while a representation such as Michelangelo's *David* is very much about a smooth, contoured, closed surface, its realism derives from its referencing of an interior beneath that externality. David seems alive and dynamic because his exterior contours suggest a depth of muscle and bone, a bodily interior that presses outwards against the external skin. Key to the artistic flowering of the period was, of course, a new understanding of the body that married art and science; this was the age of Leonardo da Vinci, whose representations of the human body were informed by first-hand experience of the body's interior gained through the dissection of cadavers. Even as the body was highlighted as an external material possession to be cultivated, there was an unprecedented interest in what lay beneath its surface, and how it functioned. In the words of Daniela Bohde, 'Knowledge of the inner body served not only to represent the human figure 'correctly', but also to enhance the status of ... art metaphysically. One was no longer compelled to illustrate the body from outside, but now comprehended the inner principles of the body's construction' (Bohde 2003: 21).

At this time, therefore, the body was being subjected to various forces that acted against inherited modes of perception. On the one hand, the perception of the body as unitary and exteriorised was being countered by a scientific focus on interiority and division, while on the other the body as interacting with and implicated in its environment and other bodies was being counteracted by a representational and social focus on the body as closed and self-contained. The body remained, of course, highly conceptualised and significant, but the conceptualisation produced by inherited modes of perception was being supplemented – perhaps even supplanted – by new techniques and conventions of looking and interacting.

The role of the anatomy theatre in Renaissance culture is of central importance here: it tied art to science; exteriority to interiority; materiality to representation. And it, like the other Renaissance themes I've briefly discussed, straddled a transformation in how the body was understood and experienced. On one hand, the dissections of Andreas Vesalius and other professors of anatomy laid the foundations for a modern, materialist and mechanistic scientific understanding of the body, and those of da Vinci and other artists informed representations of the body with a sense of interior structure; on the other, Rabelais, Bakhtin's hero of the grotesque, also conducted autopsies (Ferrari 1987: 103), and the business of dismantling a human corpse (one most likely executed for criminality and perhaps even purloined illegally) is a grotesque – not to say grisly – business. The anatomy exhibitions of the Renaissance served a variety of purposes for a variety of classes of people, from the learned men seeking insight into the body as rational mechanism to the carnival revellers seeking thrills in gory spectacle. It was a ritual space that not only changed understandings of the body through the views if offered of the bodily interior, but also by inculcating in its audience particular styles of looking and acting.

The Discovery of Anatomy

The Renaissance produced iconic representations of the body outside the boundaries of what would today be considered the world of art; indeed, the division between artistic and other forms of representation was much less final then than it is now. Perhaps the most famous representations of the body produced outside what would now be considered art during the Renaissance are the anatomical illustrations that accompany the work of the famed anatomist Andreas Vesalius. Today considered the father of modern anatomy, his seven-volume masterwork *De Humani Corporis Fabrica*, was published in the same year as *De Revolutionibus Orbum Coelestium* (Le Breton & Walker 1988: 61–2), and can be considered as important to our understanding of the human body as Copernicus's work was for our understanding of the heavens. (Indeed, at the time, the human body was still widely believed to be a divinely draughted map and microcosm of the Universe.)

Vesalius was not the first person to conduct a human dissection (even though he is credited with bringing a new approach to the practice), and in many ways he and his work seem alien to us today, too much a part of the mode of thought he helped consign to history to be fully comprehensible to a contemporary observer. Nevertheless, his work helped to transform older traditions into the dominant system of objectifying the body in our contemporary culture: medicine. And yet, centuries later, the *Fabrica's* illustrations fascinate less for their modernity than for their blending of a familiar mode of anatomical representation with an artistic sensibility and iconography that seem bizarre and misplaced to the modern reader (see Harcourt 1987: 28).

Dissections of the human body have taken place since classical times (Sawday 1995: 79–80), and the Renaissance practice of dissection was understood to be following in the footsteps of classical medicine as most famously personified in the Roman physician Galen. However, Vesalius is credited with challenging the authority of Galen by espousing the superiority of first-hand observation and investigation, rather than the unquestioning acceptance of classical wisdom (Harcourt 1987: 40).[2] First-hand observation was considered the only reliable evidence, and in this approach he is widely credited with being the first scientific investigator of the body as we would understand such a role today, and, like Copernicus, one of the founders of the modern scientific method. For Vesalius and his followers, the body should be understood by objectifying it and subjecting it to an objective gaze.

Following the cessation of such practices in the classical world, anatomical study continued in the Islamic societies, which had inherited much of its knowledge, but the European public autopsy is an idiosyncratic practice thought to have been inaugurated at the University of Bologna at the hands of Mondino de'Luzzi when he dissected two female cadavers in 1315 (Cazort 1996: 14). The public autopsy was a highly ritualised affair: Mondino had received permission from the Pope to conduct this first event, and anatomists would pay for special masses to be held concurrently with the autopsy for the benefit of the cadaver's soul (Ferrari 1987: 51). In fact, while Mondino had presumably introduced public anatomy demonstrations for the practical purpose of aiding in the training of doctors and surgeons (Ferrari 1987: 53), as the practice of public anatomisation spread through the universities of Europe this aspect came to be rendered a secondary consideration in the conduct of the event. Private dissections became the forum for hands-on medical training, while public dissections were focused on ritual and the body's philosophical significance (Klestinec 2011).

By the early fifteenth century, the public dissections at Bologna had become public spectacles held in a grand, purpose-built theatre. They took place during carnival and were attended by a cross-section of the public, a great many of whom were presumably there only for the cheap thrill of seeing a naked, dead body being cut to pieces. Tickets were sold to the event, and musicians entertained the spectators (many of whom could not understand the anatomy professor's Latin explanation of what they were seeing), while banquets were held during periods of rest in the anatomist's work.[3] The public anatomy exhibition had become a form of ritual, increasingly ill-suited for the discovery of anything new about the workings of the body as the science of medicine progressed (Ferrari 1987: 64–90, Egmond 2003: 113–14) but clearly still serving a valuable function as a spectacularisation of the body consumed by an eager audience. In the words of José Van Dijck,

---

2   Although this does not necessarily mean that Vesalius himself saw his work as constituting an attack on Galen (see Cunningham 1997: 121).

3   Cynthia Klestinec argues that the famed Bologna example has unduly influenced wider understandings of the Renaissance anatomy demonstration, given that elsewhere these demonstrations did not coincide with Carnivale and thus did not feature such a carnival atmosphere. It is reasonable to argue that the scientific and medical significance of demonstrations elsewhere was also lost on much of their audience, but not that they were universally understood as riotious, carnivalesque events.

With public dissections, anatomists did not intend to share their knowledge with a general audience, but to impress them and command respect and awe. Anatomical theaters flourished as they turned into cultural centers, where scientists and artists worked side by side, inspiring one another. Many famous painters and writers, most notably Rembrandt, attended public dissections and recorded the anatomical spectacle in their art works. Although few artists really understood the Latin oracle in front of the cadaver, they were fascinated by the combination of scientific aura, moral transformation and morbid entertainment. (van Dijck 2000: 281)

The public anatomisation was an event that served a number of functions, such as the ritualised destruction of the bodies of the criminals who supplied its cadavers and the legitimation of new kinds of knowledge and the generation of prestige for universities (Egmond & Zwijnenberg 2003: 108–20). In addition, its ceremonies established the dignity of its interference with the bodies of the dead by symbolically placing the inner mysteries of the body at the centre of the Universe as God's masterwork, whose divinely engineered mechanism served as a microcosm of and key to the organisation and proportions of the cosmos (Cunningham 1997: 41, Sawday 1995: 75). In the words of Andrew Cunningham, 'Anatomy was not a study which was primarily of value to medicine, and also about God. It was the other way about: it was primarily about God, and also of value to medicine' (Cunningham 1997: 38). However, even as this investigation of the body justified itself in these grandiose terms, the investigation of body as mechanism inevitably carried a capacity to downgrade God's role by highlighting the body's capacity to function according to straightforward, non-miraculous means. Just as Copernicus's mechanistic universe seemed able to function reliably without direct divine intervention, the mechanistically understood body lost some of its wonder and sense of divine association.

Copernicus's telescope facilitated the gathering of evidence through observation by varying and increasing the perceptive powers of the human eye, but there would seem to be a fundamental difference between studying the cosmos with a telescope (or studying microcosms with a microscope, for that matter) and seeking new discoveries through dissection. Both pursuits seek knowledge by opening up new vistas to the human eye, but where the astronomer aims to penetrate those spaces furthest from the human body, the anatomist seeks to penetrate the human body itself, perhaps on occasion even penetrating the very organ of sight. Vesalius sought new knowledge, not beyond the human scale, but rather within the human frame through a form of investigation that turns the human body inside out to scrutinise its interior, inevitably destroying it in the process. Such an endeavour was perhaps the greatest triumph of the observational method, demonstrating to those lucky enough to attend a public autopsy that such a gaze transformed what was most familiar to us – our own bodies – into what was most unfamiliar and alien: a fascinating and disturbing landscape that in other contexts was as exotic and unreachable as the surface of the moon. And it achieved this in spite of a widespread horror at its very method of operation; the cutting open and investigation of a body – even after death – remains to this day associated with superstition, religious prohibition, or simple squeamishness. Indeed, much of the ritual that accompanied the Renaissance dissection – the accompanying masses and the often elaborate burial rites for the anatomised remains, for example – were clearly intended to mitigate the negative significance of what was being done (Wilson 1987: 62–3). When it is borne in mind that Descartes was profoundly influenced by the practice of dissection, and even carried out dissections himself, the sense of alienation from his own body noted previously – the sense that his body was somehow not him, that his essential self was of a different order entirely and located elsewhere – immediately becomes more comprehensible.

Clearly, whatever innate significance the human form has for human perception is based on a fundamentally different mode of seeing. If the body is experienced and perceived as exteriority, the central purpose of anatomy is to split that exteriority open, to break apart the body as a holistically intelligible entity and dismantle it. Rather than a surface of meaning or sensation, the body becomes a collection of interlocking mechanisms that can be isolated and considered separately. It becomes a piece of engineering, which can be pulled apart and explained in technical terms.

This perhaps explains the status of the public anatomy exhibition in the Renaissance, coming as it did to be the most inspiring and fascinating example of a new way of knowing and thinking. Jonathan Sawday describes this historical milieu as a 'culture of dissection' (Sawday 1995: 2), in which the anatomy theatre and the secrets it revealed not only exerted their own, direct fascination, but also became a dominant trope for understanding human experience and contemporaneous social and intellectual changes.

> [P]artition stretched into all forms of social and intellectual life: logic, rhetoric, painting, architecture, philosophy, medicine, as well as poetry, politics, the family, and the state were all potential subjects for division. The pattern of all these different forms of division was derived from the human body ... And it is in this urge to particularize that 'Renaissance culture' can be termed the 'culture of dissection'. (Sawday 1995: 3)

From the political structure of the European state (Hobbes 1996: 9) to the unfamiliar terrain of the New World, the project of exploring and mapping the human body became representative of all manner of efforts to investigate new horizons or fix complex structures (see also Stafford 1991: 48ff.).

Sawday's 'culture of dissection' had a crucial influence on the development of the themes explored in this book at both a very particular level and a more general one. The particular level of influence is readily apparent: what the anatomists of the Renaissance saw inside the human body directly informed a new, mechanistic, understanding of the human form and the human subject. But, as indicated by dissection's wide-ranging power as a metaphor for investigation noted above, there was also a much more general sense in which it became the archetypal pursuit of the Western scientific mind. Ironically, while the pursuit of anatomy gradually eroded older understandings of the human form that saw it as a microcosm of the wider Universe and evidence of a divine blueprint, it also created a secular, scientific mindset that took the work of dissecting the body as the archetype for all pursuit of knowledge about the laws of the cosmos as a whole.

In the more particular case, anatomisation transformed bodies into fragmented, alienated objects that resisted the holistic perceptual mode and were amenable to disinterested, conscious consideration. The shock and exhilaration of seeing what lay inside the human body fundamentally changed how it was understood in Western culture. Dissection did not literally afford human beings their first glimpse of the inner workings of the human body, of course; the innumerable violences and misfortunes of human history up to that point had provided ample opportunity to see bodies in various states of evisceration and dismemberment. But dissection required the development of a particular way of seeing, one that replaced previous glimpses of catastrophic disorder with a sensible and schematised view of anatomy as carefully structured mechanism.

Cutting open a human body doesn't present to view a neatly ordered set of readily-differentiated components of the kind seen in anatomy diagrams; instead it reveals a bloody, riotous mess already in the process of corruption and disintegration. The first challenge for an anatomist like Vesalius was to train his perception so that some design could be seen in the tangle of viscera, various component parts recognised and separated from one another and then – almost as challenging – represented in a form that captured the three-dimensional promiscuity of fluids and tissues in a comprehensible way (Sawday 1995: 129ff.). In other words, the anatomist was required to learn – even create – an entirely new way of seeing the body, one that had not previously existed because it had not previously been needed. Prior to the Renaissance, the human animal had had little opportunity and still less cause to scrutinise the human interior, and so required no natural faculty for making sense of its crowded, dripping complexity. Vesalius and anatomists like him were doing more than simply documenting what was revealed by the parting of the skin: they were formulating an entirely novel way of seeing and representing the human body.

Ironically, however, this exhaustive documenting of the human body couldn't help but miss a great deal for one simple reason: the anatomist was studying a corpse, not a living human subject. After all, over eighty years would pass after the publication of Vesalius's *Fabrica* before William Harvey correctly explained the human circulatory system; the function of the human heart and blood, which seems so obvious to us today, is poorly illustrated by a corpse. More broadly, the anatomy display presents a particular view of the human body, one that lacks internal animation and dynamism but has been imbued with a strong sense of differentiated working parts available for functionalist explanation. It is not what we in our everyday lives understand a human self to be; its sense of unity and organic wholeness has been undone, and it does not move, speak, or experience emotion, among other things. It is, in other words, a vision of a machine at rest.

At the time that René Descartes was writing his *Treatise of Man* (quoted in the introduction to this book), he was living in Holland, then the centre of human dissection. Jonathan Sawday notes that, at the same time that Descartes was haunting the butchers' quarters and anatomy exhibitions of Amsterdam and Leiden (home to the most famous of all anatomy theatres[4]), the artist Rembrandt was doing the same, and in 1632 he produced his famous painting *The Anatomy Lesson of Dr Nicolaes Tulp* (Figure 2.1), in which the eponymous anatomist is shown using a set of forceps to demonstrate the working of the human hand, tugging the flexor muscles to move its fingers (Sawday 1995: 148–53).[5] The significance of this action seems clear: the anatomist is operating the body like a machine, inserting a metal probe to operate the mechanism that – during life – generates movement. Dr Tulp stands in for the absent Cartesian *cogito*, pulling the levers of the otherwise purposeless materiality of the human body, and for Descartes the significance of such demonstrations was clear:

> [T]he difference between the body of a living man and that of a dead man is just like the difference between, on the one hand, a watch or other automaton (that is, a self-moving machine) when it is wound up … and, on the other hand, the same watch or machine when it is broken and the principle of its movement ceases to be active. (Descartes 1988c: 219–20)

But the work of Vesalius and others like him had a more general impact than simply presenting the human body as a machine lacking intrinsic motivation and animation. As suggested above, the anatomical investigations of the Renaissance and later produced a new way of looking at bodies that was radically different from that employed previously, and a new mode of representation distinct from that employed at earlier times. Most striking about the illustrations of

---

4   Holland overtook Italy as the centre of anatomical research in the seventeenth century, to be surpassed in turn by London and Edinburgh in the late eighteenth (Cazort 1996: 20–23).

5   This representation is thought to be a referencing of Vesalius by way of a portrait in which the famous Belgian is shown giving a similar demonstration (Schupbach 1982: 9).

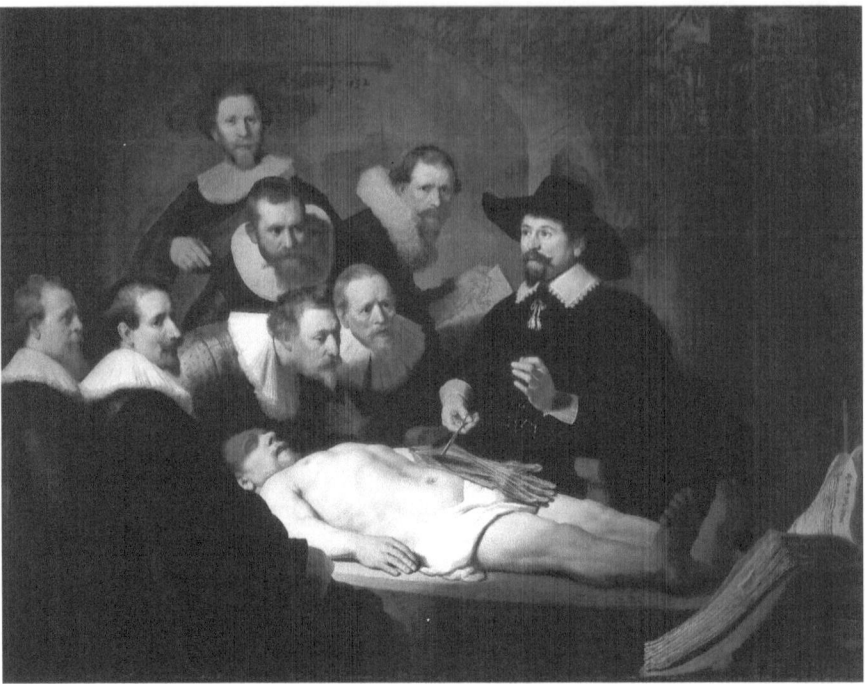

Fig. 2.1   **Rembrandt van Rijn (1606–1669),** *The Anatomy Lesson of Dr Nicolaes Tulp,* **1632. Canvas, 169.5 x 216.5 cm, inv.146. © Royal Cabinet of Paintings Mauritshuis, The Hague. Reproduced with permission.**

Vesalius's *Fabrica* is its unfamiliar conventions of representation, in which neo-classical nudes go about their business in a rustic or ruined landscape, seemingly untroubled by their trepanned heads, exposed intestines, or aprons of dangling skin (Figure 2.2).

Various readings have been put forward for these figures,[6] but Jonathan Sawday has provided a straightforward explanation for their conventions of representation: these figures appear strange to us today only because we look at them from the other side of a great divide in how the body is understood and represented. For us, the dissected body or the medical diagram should be entirely decontextualised. It requires neither a framework of art, nor religion, nor poetry to be comprehensible; in fact, the separateness and self-justification of science and medicine make any intermingling with these things seem bizarre and inappropriate. But Vesalius, as an individual engaged in the project of imbuing science and medicine with this special status, was not working in a milieu in which that special status was already available. The Renaissance public

_____

6   See, for example, Le Breton & Walker (1988: 65–6).

Fig. 2.2   Vesalian 'Muscle Man'. Image reproduced with permission
from Octavo Corp. and the Warnock Library.

autopsy had little scientific value in the sense we would understand it today, but was rather placed within a complex set of religious and social rituals and significances that have since fallen away (Sawday 1995: 75). At the same time, the interior landscapes laid bare by Vesalius could only make sense when provided with a context of wide, cosmic correspondences.

> The dissected human body in isolation signified very little because, in terms of the paradigms of natural science, the body could not yet provide its own rationale for division. It was only after the great anatomical explorations of the sixteenth and seventeenth centuries that it became possible to view the body in spatial isolation, with the implied context provided by the rigours of scientific investigation. Instead, in these earlier images, the body needed to be contextualized, juxtaposed with other motifs, shown in extravagant postures, in order to demonstrate the full human and divine significance of the anatomist's skills. It was as if the devices of art were engaged in a form of propaganda on behalf of this new discovery of the human world. (Sawday 1995: 116)

Vesalius's anatomised figures are transitional ones, laying the groundwork for a new way of seeing the body and representing information that did not yet fully exist. Not only did these representations need some measure of justifying context given the absence of a special status conferred by simple tradition, but they also needed to insure themselves against the sheer shock and scandal that might be inspired by the sight of a dismembered body (Harcourt 1987: 38).

At the same time, the technical complexity and instructional application of this nascent mode of bodily representation initiated the development of identifiably modern representational techniques. For example, Sawday highlights the introduction of the 'keyed' diagram, a new convention that allowed the attachment of dense explanation to the complex image, '"fastening" the representational body within a text' (Sawday 1995: 132, see also Sawday 2007: 87–8).

The anatomised body provides an effective illustration of the relationship between the Renaissance body as individual possession, artistic representation, and object of scientific investigation. The anatomically rendered body was open and vertically arranged; while the body as site and object of interpersonal interaction is an exterior plane, the anatomist breaks open that exterior plane and draws it aside as something of relatively little interest, burrowing through successive vertical layers of arrangement – through muscles, veins, nerves, organs and bones. The anatomical body is generalised and interchangeable; after all, no-one investigates his or her *own* body through anatomisation. Another body, ideally anonymous, stands in for the bodies of those who observe, and, as a result, the focus is on standardisation and generalisability. The anatomised body should be as 'normal' and unremarkable as possible, and its features should be generalised so that they are applicable to other bodies. Finally, this body, which is de-individuated and composed of mechanistic systems but lacking any animation or visible means to actuate those systems, seems like a machine awaiting an operator, a tool whose operator is not present. In other words, rather than *being* the individual agent, the anatomised body seems to be a possession or tool of a (now absent) individual agent, something that has been utilised to act upon the world, but which is in itself largely non-individuated and inactive without some separate locus of agency and motivation.

This new mode of understanding rendered the body more a form of property than an inalienable part of the self. It was a valuable form of property, certainly, which its owner should care for and present to its best advantage – perhaps even decorating a wall of one's home with its image – but it was private property: separate, individualised, taken out of the larger system of communal and even cosmic significances within which it had previously been located.

Vesalius's figures, wandering desolate landscapes with their bodies in various states of ruination, have not just had the ties securing their own incorporation severed, but have also been cut off from their communities, and even the cosmos (see Le Breton & Walker 1988: 65–6). Where Copernicus's *De Revolutionibus* demoted the human realm from the fulcrum of the Universe to just another lump of circulating flotsam, the work of Vesalius – whose anatomical demonstrations in his own time drew their authority from the idea that the human body was the masterwork of God and a universally applicable microcosm – facilitated a process by which the body gradually became just another fallible contraption thrown together randomly by forces of nature. This process would take some time to reach its end, admittedly, but clearly, by the time of Descartes some one hundred years after the publication of the *Fabrica*, it was already well under way. For Descartes the body might still have been a divinely wrought contrivance, but rather than the key to the Universe it was just a clever machine utilised by a soul located elsewhere.

As a tool and possession, a machine owned and cultivated by an immaterial agency, this body became something to be worked upon, maintained and utilised with maximum efficiency. We have already seen how, during the Renaissance, many individuals devoted themselves to the improvement and 'civilising' of their own bodies through the inculcation of refined sensibilities. At the same time, the spread of such refined sensibilities worked against the ritual of public autopsy. In fact, the highly structured and ritualised nature of the public anatomisation could be utilised to discipline the bodies of often rowdy and fractious university students, teaching them habits of meditative attentiveness and disinterested observation in line with the more general move towards the inculcation of 'civilised' habits at the time (Klestinec 2011: 111–22). While the autopsy theatre had straddled a cultural shift, marrying a modern project of first-hand scientific enquiry with a grotesque spectacle of gore and ritualisation of the human body's cosmic significance, the refined modern individual is unlikely to consider such a grisly spectacle a desirable public event.

Ultimately, being of questionable scientific value and distasteful to the well-mannered observer, the grotesque spectacle of the public autopsy gradually fell from favour. The period of the public anatomisation lasted roughly from 1450 to 1800, but, as its influence waned with the coming of the Enlightenment, so too was much of its role taken up by less confrontingly grisly representations of the body's workings, representations that were purged of messy biology such as the anatomical model (Egmond 2003: 108).

In addition to being inoffensive to delicate sensibilities, models had the advantage of presenting the body in a way more amenable to human understanding, free as they were from the gore and mess of a real cadaver. But the significance of the model potentially extends further. Not only could an artificial body illustrate the design of the living body in a more readily comprehensible way, but it potentially could even simulate those bodily processes lost to the cadaver, such as circulation. With the body now understood primarily as a piece of engineering, the ultimate test of the mechanistic theorisation of the body was to reverse engineer the body, building one's own model body in order to test the limits of mechanistic explanation.

The Automaton

The spectacle of public dissection came to be replaced by various presentations or representations of human anatomy that sidestepped its potentially salacious or disturbing character. There were printed anatomical references, of course, but various technologies also were employed to create three-dimensional representations of the body's interior. *Ecorchés*, or flayed figures, like the playfully posed works of Honoré Fragonard, were made by injecting skinless cadavers with wax (Landes 2007: 104), and skilled artisans catered to a burgeoning demand for anatomical models sculpted from wax (Stafford 1991: 21, 64–5). As Giovanna Ferrari notes,

> The creation of great collections and laboratories of anatomical waxworks in the eighteenth century seems in part to be attributable to the desire to substitute for the crudity of dissection a whole armoury of non-perishable pieces. These had the added advantage of being artistic euphemisms. Giuseppe Galletti, surgeon at the Arcispedale of Santa Maria Nuova in Florence, hints at this in a letter about one of his writings: 'I treat here of how, under the glorious government of the Great Leopoldo [Pietro Leopoldo di Toscana], the desire to see, without feeling nausea, and at a close range, the various parts of the body by means of wax figures, reawoke'. At roughly the same time, very similar observations prompted the decision in France to resort to models in the training of midwives. (Ferrari 1987: 105–6)

The greatest advantage of these simulated bodies was, of course, that they were clean and did not rot, and thus could be owned, kept and admired in perpetuity. The secrets of bodily interiority traded in by the public dissection could thus be owned by privileged collectors and enjoyed in private, and Ludmilla Jordanova has drawn attention to the clearly eroticised character of this consumption, manifested in the many beautiful reclining female figures amongst the models:

> It is, I think, undeniable, that the wax 'sleeping beauty' figures are knowingly erotic. At the same time they invite us to peer into bodily recesses and to find there evidence of reproductive capacities. Three distinct issues can be discerned here; the evocation of an abstract femininity, the route to knowledge as a form of looking deep into the body, and the material reproductive processes associated with women. (Jordanova 1989: 50)

Many of the themes associated with the shift from public autopsy to anatomical model were reactivated in the late twentieth century, when Gunther von Hagens reintroduced the écorché through the 'plastinated' cadavers of his Body Worlds exhibitions (Anon. 2006). Von Hagens explicitly positioned himself in the tradition of Vesalius, and cast himself as glorifying the human body and enlightening the public by providing them with access to a view of the body usually reserved for a specialist group (Hirschauer 2006: 31). In so doing, he resurrected an unease concerning grave robbing, grotesquery and public display of the bodily interior that more or less picked up where it had left off with the end of the age of the public autopsy. However, while von Hagens's plastinates are made from real bodies, they are in many ways more like the wax figures that replaced most autopsied cadavers in their sanitised and 'eternalised' state. In fact, given that the plastination process entails removing all water and fat from the cadavers and replacing it with plastic (see Hirschauer 2006: 28–9), it is in fact the case that almost all the original material that makes up a living body has been removed, and its wet, decaying biology has been sanitised. In the words of Tony Walter, 'By transforming the body's slimy, slippery, smelly interior into dry, odourless, colourful exhibits, plastination produces an anatomy that is acceptable to civilised sensibility' (Walter 2004: 478). While von Hagens's work seeks to revisit a prior moment in the history of the human body's relationship to art and science, therefore, it still does so by using new technologies to turn the body into a sanitised, diagrammatic artefact.

The technologies available for the creation of écorchés and models was less sophisticated in centuries past, of course. Nonetheless, as the era of public dissection came to an end, there were a number of notable attempts to realistically model the attributes of human bodies using what was the cutting-edge technology of the day: clockwork.

The clockwork automaton was a highly ambitious attempt to represent the workings of the body. Being a machine, it is free from the stench of corruption and the leak of fluids, the messy promiscuity of tissues and frustrating idiosyncrasies of actual bodies, and yet it can provide the animation missing from both cadavers and wax models. It is clean, neat and reliable, and subscription to mechanistic explanations of the body makes machinery seem like the most appropriate medium for its simulation. However, it does introduce a circular logic: a belief that bodies are like machines motivates the simulation of the body using machinery, but using a mechanical simulation to investigate bodily processes cannot help but generate a view of the body that further entrenches the idea that it is machine-like in nature. Like any representation, it transforms the object to which it makes its appeal (cf. Landes 2007: 102–3).

Automata, or 'self-moving' artefacts, have existed since classical Greek times, and during the mediaeval period Islamic mechanists had inherited the Greek tradition, while simpler 'clock Jacks' and animated religious figures had been built in Europe (Hillier 1988: 14–19). This tradition continued during the Renaissance – Leonardo da Vinci reputedly built a mechanical lion, which reared in the air to celebrate Francis I's visit to Lyon in 1515, and his designs even include a blueprint for what seems to be a mechanical knight (Moran 2006: 989–90). By the eighteenth century, advances in the production of clockwork had provided more sophisticated mechanisms for automata (Shieber 2004: 17–19), but, more importantly, the significance attributed to them had also changed. They were now understood to be

> philosophical experiments, attempts to discern which aspects of living creatures could be reproduced in machinery, and to what degree, and what such reproductions might reveal about their natural subjects. Of course, … automata were also commercial ventures intended to entertain and demonstrate mechanical ingenuity. But their value as amusements lay principally in their dramatization of a philosophical problem that preoccupied audiences of workers, philosophers, and kings: the problem of whether human and animal functions were essentially mechanical. (Riskin 2003b: 601)

The automata of past ages had been designed simply to divertingly reproduce the appearance of living things, and as a result the hidden mechanisms by which they did so were of little concern to the observer; the illusion of life was enough. However, in the eighteenth century, the influence of mechanical philosophy had introduced a desire for automata to be actual simulations, rather than mere representations, created to 'reproduce physiology' (Riskin 2003b: 602).

We have already seen how the technical developments of the anatomy theatre informed the mechanical philosophy of Descartes. But for Descartes the connection between machine and body suggested by anatomical studies was more than a vague metaphor or thought experiment; I have already quoted him making quite explicit connections between the human body and automata, but even before he expressed this connection in philosophical terms he considered it in practical terms, developing a plan to create a reproduction of the human body animated by magnets. In 1619 he reportedly developed plans to build various automata: a dancing man, a flying pigeon, and a spaniel that chased a pheasant. There is even an apocryphal story about Descartes building a mechanical copy of his illegitimate daughter Francine, who had died of scarlet fever at the age of five, which came to an unfortunate end when the crew of a ship on which Descartes was travelling discovered her and, seized with superstitious dread, threw her overboard (Maisano 2007: 63, de Solla Price 1964: 23, Wood 2002: 3–5).

But by the eighteenth century, the mechanistic understanding of the body had changed in an important way. Descartes had argued for a *machina carnis*, a machine body, but his understanding of human beings was, of course, fundamentally dualist in nature. He might have hoped to create a mechanical body, but he did not believe it was possible to create a mechanical human being because the human soul was not reducible to mechanism. For him, the use of language represented a division between man and even the cleverest machine – not to mention animals, which in his view *were* ultimately no more than mechanism (Gunderson 1964: 198).

As we have already seen, it was in the eighteenth century that the doctor Julien Offray de La Mettrie[7] most famously and scandalously argued that human beings were *nothing but* mechanism, and that Descartes' soul and *res cogitans* were illusions, nothing more than an epiphenomenon of the workings of '*l'homme machine*': 'Given the least principle of motion, animated bodies will have all that is necessary for moving, feeling, thinking, repenting, or in a word for conducting themselves in the physical realm, and in the moral realm which depends upon it' (La Mettrie 1927: 48–9). For La Mettrie and those who shared his views, Descartes had been heading in the right direction, but hadn't gone far enough. La Mettrie therefore excuses Descartes as a product of his age even as he ridicules his belief in a soul:

> I believe that Descartes would be a man in every way worthy of respect, if, born in a century that he had not been obliged to enlighten, he had known the value of experiment and observation, and the danger of cutting loose from them … This celebrated philosopher, it is true, was much deceived, and no one denies that. But at any rate he understood animal nature, he was the first to prove completely that animals are pure machines. And after a discovery of this importance demanding so much sagacity, how can we without ingratitude fail to pardon all his errors! (La Mettrie 1927: 63)

---

7   For a more detailed account of the transition from Descartes' ideas to La Mettrie's, see Channell 1991, Chapter 3.

La Mettrie therefore understands himself to stand on the far side of a bridge in scientific thinking that Descartes helped to build, but could not himself cross. The mechanism of the Enlightenment is different in fundamental ways from the mechanism that preceded it, and the perceived significance of automata is also different as a result. In the words of Jessica Wolfe,

> While there indeed exist 'machine-men' before the rise of the corpuscular philosophy, they do not necessarily behave in ways sympathetic to a post-Cartesian understanding of what it means for a human being or an animal to be 'like a machine' or, for that matter, what it means for a machine to be lifelike. One of the profoundest yet most imperceptible legacies of the intellectual and scientific revolutions of the seventeenth century is the way in which new systems of classification and discourses of knowledge alter the perceived relationship between humans and machines. (Wolfe 2004: 240–41)

La Mettrie's materialism was far from uncontroversial even in his own time, and he was twice forced to relocate in order to escape the hostility generated by his writings.[8] Nevertheless, the power of such materialist ideas was clearly to a substantial degree both a result of, and a motivation for, the more complex clockwork automata being constructed at the time. Most famous of the eighteenth-century builders of automata was Jacques de Vaucanson, who achieved fame and wealth by exhibiting his creations: a flautist that could play twelve melodies (created in 1735), followed by a second musician that could play twenty melodies, a mechanical duck, and lastly another musician with a repertoire of twenty melodies (Moran 2007a: 679).

Jessica Riskin has called attention to the danger of drawing too simplistic a connection between the work of Vaucanson and other automata makers[9] and the mechanism of La Mettrie (Riskin 2003b: 610–11). At the time Vaucanson was securing his fame, La Mettrie's *L'homme machine* had not yet even been written, and the mechanist account of the human body was not in the ascendant. Indeed, there is a stark contrast between the tremendous popularity of Vaucanson's automata and the hostility with which many greeted La Mettrie's theories. Riskin's account suggests that automaton makers like Vaucanson were simply testing the limits of the mechanistic explanation of

---

8   La Mettrie was an abrasive personality, and the hostility he provoked in others was not only caused by his mechanist views. Nonetheless, when he published *L'homme machine* in Leiden, the famous centre of anatomy, the church ordered all copies of it burnt and La Mettrie was forced to flee the country.

9   It is not necessary for me to provide a more extensive list of automaton makers or their creations here, as I am only pointing to the interest in such artefacts at the time, not the specific nature of individual mechanisms. For detailed discussions of automata makers such as Vaucanson, Jaquet-Droz and von Kempelen, as well as their automata, see, for example, Bedini 1964, Riskin 2003b, Hillier 1988: 44–58.

Fig. 2.3 A 1738 representation of Jacques de Vaucanson demonstrating his automata. © Mary Evans Picture Library. Reproduced with permission.

human life, rather than seeking to prove their unadulterated usefulness. In fact, Vaucanson's most famous creation, the 'defecating duck', like many other similar devices, was partly showmanship, only creating the illusion of reproducing animal physiology rather than proving that it could be recreated by machinery (Riskin 2003b: 609–12). Nevertheless, Vaucanson clearly understood his creations to be generating insights into the nature of living bodies. Importantly, Vaucanson's motivation in building his famous automata seemed to be primarily the investigation through simulation of anatomy and physiological processes. As a young man, Vaucanson had studied anatomy, and at twenty-two had met Claude-Nicolas Le Cat, a French surgeon who had written a treatise entitled *Déscription d'un homme automate* (*Description of a Human Automaton*) (Landes 2007: 100–101).[10] While Vaucanson achieved fame through the automata he did build, throughout his career his planned masterwork had been a mechanical 'moving anatomy' that could simulate in detail the internal processes of the human body, a wonder he vied with Le Cat to create (Wood 2002: 50, Riskin 2003a: 114), but, despite support from Louis XV, this dream was never realised (Riskin 2003a: 601). In the words of Joan B. Landes,

> For Vaucanson, the conventional approach to anatomical knowledge – earned through careful empirical observation of cadavers and dissection – was by itself inadequate and ought to be supplemented by a visual demonstration of a working model or lifelike mechanical simulation. Motivated by a strong interest in anatomy, the engineer's inquiries were also intended as contributions to anatomy, designed to further the comparative science of animal and human bodily structure ... (Landes 2007: 98)

Whatever the truth of such automata's workings or the intentions of their builders, their audiences, inspecting these ever more cleverly designed mechanical bodies, might have been forgiven for believing that they established the value of functionalist accounts of the human body that saw it as merely a complex machine (see de Panafieu 1984: 130–31). Certainly it is clear from La Mettrie's writings that he considered them to support his cause, and the increasing complexity of humanoid machines suggested that it would one day be possible to simulate all aspects of human biology:

> [Man] is to the ape, and to the most intelligent animals, as the planetary pendulum of Huygens is to a watch of Julien Leroy. More instruments, more wheels and more springs were necessary to mark the movements of the planets than to mark or strike the hours; and Vaucanson, who needed more skill for making his flute player than for making his duck, would have needed still more to make a talking man, a mechanism no longer to be regarded as impossible, especially in the hands of another Prometheus. (La Mettrie 1927: 61)

---

10   Le Cat's lost work prefigured La Mettrie's later *L'homme machine*.

The problem with La Mettrie's claims is that they use a very limited simulation of bodily action as proof of the possibility of a complete simulation of bodily action, a problem that reappears in later mechanist hypotheses based on computationalism (see Chapter 4). The automaton's limited referencing of the human body invites a belief that every human capacity might be simulated in the same way, and thus that the human body itself is not qualitatively different from the automaton, being superior only in its greater complexity. While Descartes did not accept such an idea, and even Vaucanson most likely did not either, the philosophical speculations and mechanical demonstrations of these men laid the groundwork for the claim that the machine and the human body are not simply analogous to one another, but are in fact one and the same thing (de Panafieu 1984: 131). For La Mettrie, clockwork automata demonstrated that the human body is itself simply a complex machine, and thus is not in any way opposed to other machines or privileged in relation to them. However, while the creation of automata is a key step in the development of a purely materialist and mechanist understanding of the human body, it would be wrong to believe that such demonstrations of technology will inevitably produce such an idea. The appearance of such a belief is reliant on a wider context of thought, and this lack of inevitability is evidenced by the Japanese *karakuri* tradition.

Reaching its apex in the nineteenth century, the Japanese *karakuri* was a self-animated toy similar to the European automaton, one which also largely took human form and enjoyed the same degree of popular appeal (Screech 1996). Initially animated by means such as ropes and weights, the introduction of clockwork from Europe led to the creation of more complex automata that independently paralleled their European counterparts.

When the first clocks arrived in Japan in the sixteenth century, the country was under shōgunal rule and pursuing a policy of *sakoku* or 'closed country'. European clockwork was seen as a key area of Western technological superiority, and the use of elaborate clocks as gifts was a valuable part of Western diplomatic strategy (Bedini 1964: 40, Hillier 1988: 37–43).[11] Although a steady trickle of Western ideas and artefacts did enter the country, this only served to tantalise the Japanese public with glimpses of occidental exotica that only the wealthy and well-connected could ever hope to own, and the military and industrial application of these foreign technologies was outlawed.

Masao Yamaguchi (2002: 72–3) has argued that the diversion of technologies like clockwork into the realm of spectacle and entertainment rather than more practical applications, which resulted from this restriction, has had an important influence on Japanese popular attitudes towards technology. Where the new technologies that accompanied the Enlightenment were in the West associated

---

11 It should also be pointed out that, in Japan, the study of anatomy was also considered a key area of Western superiority, and Western anatomical textbooks were prized objects of study (Screech 1996: 87–9).

with the upheavals and dislocations of industrialisation (and, in the current context, we might add profound and troubling changes in the perceived status of the human body), in Japan they were stripped of any intimation of larger significance or threat.

> In Japan, it was in the world of entertainment, in a ludic ambience, that the automaton puppet made its appearance. As playthings, the puppets did not threaten human competence or existence; they remained charming copies of the human figure. They were considered in a way like domestic animals, or strange animals such as elephants that had been made objects of curiosity in the milieu of spectacle ... (Yamaguchi 2002: 79)

Of course, the European clockwork automaton also lacked any productive use, but it was positioned within discussions of the living body as mechanism, and its technological underpinnings were quickly assimilated into industrial processes. This humanisation of machines and mechanisation of labour brought living body and machine together in powerful ways, creating a context that was absent in Japan, where the *karakuri* could remain an innocent diversion.

Japan has its own particular history of interactions with new technologies, from which many of the crucial influences on Western conceptions of the relationship between the body and machine are absent (although, of course, those of more recent vintage are largely shared). The Japanese example is neither wholly the same nor wholly different from Western examples, and the Japanese are regularly characterised as having a more enthusiastic – even affectionate – relationship with new technologies than is present in many in Western countries.

I will return to Japan in the following chapter; however, what is significance about the Japanese case here is that, while sophisticated clockwork bodies were a prominent part of Japanese popular culture at about the same time as the European fascination with automata, without a wider and complementary intellectual context they did not initiate a process whereby 'people's assumptions about what is essential to life and what is within the purview of machinery have continually transformed each other' (Riskin 2003a: 115) as European automata did. *Karakuri* did not produce a Japanese La Mettrie, who used them to support the claim that human bodies are themselves a kind of machine.

Clearly, the association between machines and human bodies is not an inevitable result of our innate propensity to search our environment for the kinds of communicative information produced by human bodies. Vaucanson's clockwork automata were specific to a particular historical moment; they presumably would not have appeared but for the dissemination and development of ideas from mechanical philosophy, and Jessica Riskin characterises them as part of a project to realistically simulate life whose failure to satisfactorily realise it aims resulted in a further, subsequent change in how automata were understood (Riskin 2003a: 99–101). M. Norton Wise points out that, while automata of the eighteenth century were male or female in roughly equal measure, in the nineteenth they were mostly female and, '[w]hen they were not female, they were typically other sorts of uncanny or exotic creatures: talented children, blacks, acrobats, monkeys, magicians, and others' (Wise 2007: 163). This could be interpreted as reflecting a change in the status of the automaton caused by a degree of disappointment with its simulation of human life: not being able to recreate the human body as perfectly as some initially had hoped, it was seen as more appropriate for it to simulate those Others – 'women, blacks, apes' – who were understood to be closer to the realm of Descartes' *bêtes machines* than White men, and who were popularly associated less with thought and creative expression, and more with the kind of repetitive, rhythmic actions at which clockwork excelled.

If this change signalled disillusionment with the effort to reproduce certain attributes of the living body using machines, it also pointed to the efficacy of reproducing others. While the limited capacities and repetitious behaviour of machines might have made them seem like poor imitations of living bodies once their novelty had worn off, with industrialisation this perceived gap could be narrowed, not simply by the creation of new machines that could do more, but also by practices that required humans to do less. Increasingly sophisticated machines entered factories alongside workers whose independence and range of movement was increasingly restricted. In addition to his entertaining automata, Jacques de Vaucanson used similar technology to create the world's first automated loom (Riskin 2003b: 627), the ancestor of the Jacquard loom that in time would come to replace the real labouring bodies of skilled workers in France's textile industry. His automata therefore helped to usher in the Industrial Revolution, which would fundamentally change how the relationship between human body and machine was understood.

Unification and Fragmentation

The historical developments discussed in this chapter begin with the attempt to create a new way of seeing the body, a new way seen as most productive for pursuing a new mode of investigation. However, this new mode of seeing inevitably creates its own kinds of visual data – visual data that then further inflect understandings of the body. Comparatively rapidly, this new understanding in turn informs attempts to simulate bodily processes, creating a situation where this mode of understanding reinforces itself by, not only colouring perception, but also producing new artefacts and representations of the body that seem to confirm its veracity. This shift in perception broadly moves towards the mode of perception suggested at the opening of this chapter; that is, it seeks to short-circuit the holistic, exteriorised mode of perception generated by our preconscious processing of bodies, and instead transform the body as object of sight into something interiorised, fragmented and alien, a collection of strange mechanisms.

The fragmented, reductionist view of the body has an air of (literally or figuratively) breaking open its occluding surface planes and dispelling the glamour of the human form so that it can be scrutinised with a dispassionate eye. In order to counteract the special power the human form has over human perception, it seeks to transform the body into something sub-divisible; composed of more or less self-contained, separable parts and systems; separable from its environment; fixed and stable; and generalisable.

The idea that the body is a self-contained system composed of numerous subsystems that can be investigated in isolation – 'a neat and exact assemblage of related structures' (Vartanian 1973: 135) – forms the basis of the mechanistic approach, and makes possible (and results from) the science of anatomy. However, at the same time that anatomical investigations depict the body as a decomposable collection of fragmentary mechanisms, neo-classical artistic representations of the body depict it as whole and integrated, smooth and unified. Realist artistic representations of the body seek to create a visual stimulus that we perceive as sharing the qualities of real living bodies while at the same time inevitably fixing and conceptualising, stabilising and unifying the human body to a degree not possible with living bodies.

But these two modes of representation that attained dominance during the Renaissance are not antithetical in the way that they might at first appear. Artistic representations and medical representations are implicated in one another: the anatomised body is represented through art, and art uses the anatomised body as a point of reference. The whole, unified neo-classical body gains its sense of integration and solidity through the sense of an interior that has been mapped out through the practice of anatomisation, and the body's beauty comes from its harmonious integration of internal mechanisms and systems, mechanisms and systems that themselves are evidence of divine engineering. Furthermore, the unified, individualised body becomes a subsystem within a larger mechanism – the machine of society – just as the individualised body itself is understood as a collection of subsystems, continuing the idea that the body is a microcosm for larger systems of organisation.

The internal components of bodies become the concern of science and medicine, while the gathering together of these components beneath a pleasing exterior becomes the business of aesthetics, whether it be the aesthetics of art or those of self-cultivation, where an individual seeks to present a pleasing body to others rather than one that is grotesque or distasteful. Science and medicine work to escape the pre-consciously formulated perception of the human body, while art and social performance of course don't; the latter pair seek to play upon and heighten our fascination with unified bodies. At the same time, however, they do this by creating contrived (re)presentations and manifestations of bodily forms dependent on techniques of objective observation such as anatomisation and simulated perspective.

While Andreas Vesalius is famous for the way in which he dismantled actual dead bodies, this fame most importantly arises from the ways in which his interactions with cadavers were converted into textual and – most importantly – visual forms. His cadavers were fragile, disorganised, and prey to corruption; it was the ways in which he and his unknown artistic accomplice[12] transformed them into images that guaranteed his place in history and played a key role in transforming how the human body was more widely understood.

The anatomical diagram transforms a body from something individualised, messy and resistant to visual scrutiny or description, into something generalised, abstract, organised and amenable to the gaze. It makes the components of the body seem more distinct, easily isolated and understood as discrete mechanisms performing individual functions. In Rembrandt's image of Nicolaes Tulp, the body of the cadaver seems like quite a simple mechanism: pull on the right tendons, and its hand moves. Of course, the reality of what causes the hand of a living body to move is vastly more complex than that and can't simply be isolated in a small area of human anatomy; the anatomical diagram transforms the body into a set of clearly defined fragments and mechanisms, and this in turn enables a mechanistic, functionalist account of what the body is. But this is just one, initial phase in a process that has rendered understandings of the human body increasingly abstract throughout the history of Western modernity.

Once the human body is schematised as a set of mechanisms, it is rendered as simply a set of material systems, God's own engineering problem as suggested by Descartes' *Treatise of Man* (1972). And, once this happens, human engineers can set about studying those systems and creating equivalent mechanisms. We then have Vaucanson's clever automata, which themselves answer the question of how, for instance, a flautist creates music using breath and body. But, right from the beginning, the idea that the human body is a work of engineering creates ambiguity and reciprocity between the realms of technology and biology. Vesalius's diagrams, and those of countless other anatomists, represent the *actual* human body, not a machine, and Vaucanson's automata, while no-one would consider them replacements for real, living bodies, nevertheless create the impression that, as they mimic the workings of real bodies, there is a fundamental commonality uniting the engineering of nature and the engineering of human industry. Descartes' thought experiment asked us to imagine a divine engineer building the body of a man, making the human organism seem much the same as Vaucanson's creations, but even after scientists dispensed with the idea of God, the idea that living bodies were mechanisms engineered by a depersonalised nature continued to hold sway. The replacement of the Enlightenment's 'divine

---

12 See Kemp (1970) for a discussion of the identity of the muscle men artist.

watchmaker' with the 'blind watchmaker' of Darwinism didn't entirely escape the suggestion of some agency behind the design of the body.

The mechanistic view of the body produced by anatomy made possible the development of a powerful reciprocity between understandings of the body and understandings of actual machines. This reciprocal relationship certainly didn't appear from nowhere at the dawn of Western modernity – it has been posited in numerous different cultural and historical contexts – but it was an important factor in the rise of modern scientific inquiry. The fragmentation and mechanistic explication of the human body through anatomy becomes exemplary of rational, scientific inquiry in general and, despite its contribution to the gradual abandonment of mystical accounts of cosmological phenomena, it ties together body and heavens through a belief in common mechanistic principles at work in the movement of both.

Mechanism develops through a reciprocal relationship between actual machines and natural phenomena: clocks are created to accurately measure time according to time-keeping systems based upon cosmological movements, but this correlation in turn leads to the understanding of cosmological movements as being like clockwork; mechanisms are created to replace human labour, but the similarity between those mechanisms and human action then leads to the human body being explained through analogy with such machines. Today, the computer, a device designed to replace certain kinds of mental labour, provides the dominant framework for explaining mental processes themselves. The creation of this relationship, whereby mechanisms created through analogies with natural phenomena become naturalised to such a degree that the analogies that produced them are inverted, is crucial to the way in which the relationship between human body and technology has developed throughout modernity. Mechanical philosophers such as Descartes see bellows as a model for how the lungs work and pulleys as a model for the muscles, forgetting that bellows were created to reproduce the function of the lungs and systems such as pulleys were created to assist the action of the muscles. It is forgotten that such mechanisms were designed to mimic bodily organs and systems, with the result that their similarity to bodily organs and systems is taken as evidence of some larger cosmic principle that is manifested in unrelated contexts.

The circular analogies continue and reinforce themselves: not only does a mechanistic view of the body invite methods of research and explanation that support the mechanistic view (such as breaking the body down into a series of more or less discrete mechanistic systems and seeking to explain the engineering principles behind their functioning), but it also establishes a close connection between actual machines and human bodies. Machines are made to replace human action, either in production or (as in the case of clockwork automata) for entertainment or philosophical inquiry. Furthermore, the quality of automatism, or self-movement, becomes more generally associated with life, and human life in particular, so that later mechanical wonders such as the steam train or car – mechanisms with little surface similarity to the human body – become anthropomorphised too. As industrialisation brings machines and human bodies into closer relationships, and machines more and more come to reproduce and replace human physical action, this can only intensify.

At the same time, however, the mechanistic view of the human body produces important changes in how the human body functions as an object of human perception. The body as it appears to embodied human perception is a single entity that nevertheless functions as an interface with its environment and other bodies. It is a site of exchange and interaction, appearing to human vision as a collection of exterior planes and contours from which can be read an array of complex and subtle signals. But these two key attributes – the body as exteriority and the body as site of interaction and exchange – are not amenable to the mechanist approach. First, the mechanistic view treats the body's exterior planes largely as an impediment to vision and understanding, a veil drawn across the inner mechanisms with which it is concerned. The very fact that human perception tends to see bodies as unified and exteriorised, according to the mechanist approach, illustrates the superiority of the reductionist scientific method, which produces a more disinterested, technical view of the human form. Second, the mechanist approach requires that the body be understood as a closed, self-contained system composed of numerous self-contained subsystems. Consequently, the body's permeability and status as a site of exchange tends to be ignored. The mechanistic view has traditionally approached the body as a closed system that can be isolated and explained according to discrete internal subsystems, each of those subsystems in turn being understood by isolating and explaining them as further discrete (although interlocking) systems.

As a result of this, mechanism defines itself against the 'human' view of the human body. According to the human perspective on the body, the body is immanent, indivisible and exteriorised. Because this perspective on the body is largely innate to human beings, the mechanistic view requires the invention of new ways of looking and the formulation of new perspectives. This project is inevitably dependent upon technology for two reasons. First, and most obvious, new techniques are required to open up the kinds of vistas required; the internal organs, the microscopic cells, and the movement of blood, for example, can only be seen using various mechanical aids. However, there is a second advantage that comes with the technologised view, one tied to the first: with the machinic view comes the hope of an escape from an invested, bodycentric human perspective. While the cognitive equipment of human perception is customised to see the body in particular ways, the gaze of the machine – whether it be the microscope, the X-ray film or the MRI-derived computer image – bears no such investment. Of course, a human technician must then look at and interpret the perspective of the machine eye, but the technician can only hope to cultivate the kind of objective, disinterested perspective produced by the machine, and gazing upon the representation produced by the machine as well as or perhaps even instead of the living body upon which its representation is derived can make this seem more attainable. In short, a scientific account of the body needs to 'crack open' the body and scrutinise it in a way alienated from a naturalistic human perception of the body. In order to do this, the body must be fragmented and alienated from itself so that it can be studied as a dehumanised set of discrete systems and mechanisms purged of its usual significance for our innate apparatus of human perception. This can be achieved both by creating new, inhuman technologies of vision, and cultivating a kind of perception and understanding that is itself as close to inhuman as possible.

Chapter 3
Android Dreams

Trading Places

While industrialisation brought machines that (broadly speaking) could have been expected to serve more utilitarian ends than the clockwork automaton that had once held such fascination, the human form was certainly not exorcised from such technologically fabricated artefacts. The idea of a machine that could mimic the human body continued to stalk human dreams and nightmares, although the courtly entertainments of the automaton were transformed into the stolid strength and efficiency of the robot, an entity whose name, rooted in the Czech word for drudgery, reflected the new emphasis on productivity and power. No longer imagined to be a leisured class of scribes and musicians or a direct reproduction of living anatomical processes, these mechanical bodies were to labour as tirelessly and productively as the other marvels of the Industrial Age.

The robot brought with it a potential to carry body–machine analogies into more disturbing terrain. This new terrain is reflected in the way the *Oxford English Dictionary*'s first definition for the word 'robot' splits into two parts: first, a human-like machine, and second, a machine-like human.[1] Where the automaton had ultimately failed to deliver on the promise of faithfully reproducing the living body, the robot appeared at a time when the attributes of living bodies increasingly were being fragmented, isolated and reduced to mechanistic quantification, creating spheres of activity and evaluative criteria in which machines could compete with, and even surpass, human beings.

While popular expectations and fears for a new breed of machine body tended to outstrip anything human industry could actually deliver (as a result confining robots largely to science fiction films and novels), the power of machines to evoke the human form could still be seen in more subtle ways in the Industrial Age, most clearly in the streamlining aesthetic of the 1920s and 1930s.

And yet, rather than reflecting a conflation of body and machine, the very fear and hostility implicit in many fictitious accounts of robots might be taken as evidence of a fundamental opposition between them. Fantasies such as the *Terminator* films depict a literal war between human beings and technological

---

1   1a is 'An intelligent artificial being typically made of metal and resembling in some way a human or other animal', while 1b is '*fig*. A person who acts mechanically or without emotion'. The complexities of the term's definition will be discussed in more detail later in this chapter.

artefacts in which machines seek to exterminate the human race. This hardly seems evocative of a harmonious blending of machines into our conception of our own bodies; but such depictions do evoke a sense of interchangeability. Fears of machines insinuating themselves into our society and even bodies reflect anxieties about a potential inability to differentiate body from machine, and the apocalyptic future of the *Terminator* films revolves around the idea that machines might simply replace our bodies, mimicking us so well that we can't tell the difference, and ultimately populating the Earth in our place after we have been eradicated.

Samuel Butler's *Erewhon* (originally published in 1872), the satirical story of a hidden civilisation in New Zealand that has rejected modern technology, hinges on Butler's musings regarding the relationship between Darwinian evolutionary theory and the development of machines, producing the idea that machines are like organisms following their own parallel development:

> Take the watch, for example; examine its beautiful structure; observe the intelligent play of the minute members which compose it: yet this little creature is but a development of the cumbrous clocks that preceded it; it is no deterioration from them. A day may come when clocks, which certainly at the present time are not diminishing in bulk, will be superseded owing to the universal use of watches, in which case they will become as extinct as ichthyosauri, while the watch, whose tendency has for some years been to decrease in size rather than the contrary, will remain the only existing type of an extinct race. (Butler 1970: 202–3)

Drawing on previously published essays, the fictional Erewhonian sacred text *The Book of Machines* warns of the dangers of allowing machines to evolve to the point where they develop a 'mechanical consciousness' (Butler 1970: 199) and challenge human control. Butler suggests that nineteenth-century humanity is already 'a sort of parasite upon the machines'; 'An affectionate machine-tickling aphid' doomed to 'become extinct in six weeks' (Butler 1970: 205–7) were its machines to be taken away.

As seen in the previous chapter, the blurring of the boundary between machine and body has a long history, but at the same time results from very specific historical influences. While it might be assumed that mechanistic accounts of the body simply result from the fact that, in our mechanised society, the machine has become an overpowering metaphor for just about everything, on closer examination the relationship between machine and body is more complex. It is not at all clear that one idea is dominant over the other; it is just as common to imagine machines as bodies as it is to imagine bodies as machines, and there is clearly a great deal of exchange going on between the two concepts. Indeed, in 1964, Derek J. de Solla Price argued that mechanism doesn't spring simply from the drawing of analogies between machines and the natural world, but rather results from an innate human desire to create artificial reproductions of the natural world, which drives us to create artefacts that cannot help but invite comparison with the natural world they reproduce (de Solla Price 1964). Whether or not this argument is entirely true, what seems clear is the extent to which our society is just as interested in understanding machines as bodies as it is in understanding bodies as machines. One only has to cast an eye back over the rich history of robots in film to gain a sense of how great a fascination the idea of a technologically fabricated body has for us. But this fascination is only the most obvious example of a much deeper imaginative link between bodies and machines. Of course, this very tendency to link bodies and machines can draw attention to differences between the two: Masahiro Mori's widely cited (but not terribly helpful) principle of the 'uncanny valley' (Mori 1970), which claims that, up to a point, there is a decrease in our level of identification with simulated bodies as their level of realism increases, results from the fact that high but imperfect levels of similitude actually interfere with our ability to project bodily attributes onto non-bodies. A crude stick figure or smiley-face can strongly evoke a human body for us, but a more detailed representation whose features are slightly wrong troubles us. The human perceptual apparatus, always seeking to identify and analyse human faces, is perfectly capable of reading emotion into the rudimentary representation, but the more detailed but imperfect representation provides its own additional information that does not mesh with the brain's nuanced and complex frameworks for understanding human features. Watching the robot character C-3PO in *Star Wars*, we have no trouble attributing the figure with human emotions despite its complete facial immobility, but it would be more difficult to identify with a figure whose rubber face imperfectly mimicked human expression, and we might even find such a figure disturbing or repulsive. However, this only highlights a conflict between attempts to realistically simulate bodies and our more general tendency to anthropomorphise all manner of machines. In fact, those who seek to prepare us for a supposedly imminent robotopia are working to alleviate this problem by creating robots that make more effective appeals to our anthropomorphising tendencies. A recent study placed a Sony QRIO android in a University of California early childhood education centre in order to study children's interaction with it, working on the premise that children might be less set in their attitudes than adults. When the robot fell down, children helped it back to its feet even after being instructed not to, and when it shut down to recharge its batteries they covered it with a blanket (Tanaka et al. 2007: 17956). Perhaps the widespread fascination with humanoid robots is actually a natural inclination that has been weakened in adults by the learning of an instrumentalist attitude towards machines, and children are actually more, rather than less, likely to identify androids as human bodies. In any event, the very existence of such research into human–robot interaction is telling in that it clearly reflects a felt need to create better relationships between human beings and robots. In contrast, if the introduction of personal computers had been met with reservations by the public, presumably the computer industry simply would have turned its attention to other products, rather than funding new research into how to make people like computers. Research into human–robot interaction reflects an assumption that a robotic future is inevitable; if the general public doesn't like it, then research must shift towards developing ways of changing people so that they do.

Clearly, the association of machine and body can generate a great deal of ambivalence. While machines have, in a powerful and fundamental way, come to be associated with the human body, this association brings with it anxieties about the potential loss of human uniqueness, and a fear that we will lose the capacity to define humanity in an exclusionary way. Even as our fascination with the machine as body spurs both material attempts to recreate the body using technology and conceptual manoeuvres that seek to understand the body as a kind of machine, we are on some level terrified by the new artefacts and concepts produced by these endeavours. Perhaps nothing illustrates the power of this association more than the fact that our recurring sense of disorientation or outright terror upon producing these new artefacts and concepts has done nothing to lessen the zeal with which we go about producing them.

This chapter will consider some of the dreams and expectations surrounding the figure of the robot, basing them in an account of how they reflect the evolution of the body–machine relationship in the Industrial Age. This evolution progressed along two parallel and mutually informing streams. First, medical and scientific techniques of investigating the body moved on from those already seen in earlier anatomy, both refining the mechanistic account and seeking to improve upon it using both new investigative machines. Second, the new machines and mechanical processes of the time provided new mechanistic models that could inform understandings of what bodies were and how they functioned. The automated factory provided a key environment within which these streams converged to produce the robot, although this powerfully evocative mechanical figure has gone on to claim a significance that extends far beyond industrial labour.

Chronodissection

In the eighteenth century, attempts had been made to investigate the human body by building machines that reproduced various aspects of its life and activity – for example Vaucanson's flautist, which produced music using the same actions and movements as a human player, or his proposed anatomical model that could simulate circulation and respiration. After the initial excitement, however, these efforts were seemingly found wanting, and in the nineteenth century automata largely returned to being entertainments and diversions rather than philosophical experiments. The practice of anatomy, too, lost some of its glamour, and the disillusionment with these practices, both of which I have already argued were committed to the investigation of the body as a mechanical system, can be tied to a common cause. In the Industrial Age, energy and dynamism were of greatest concern, and neither dead bodies nor wind-up mechanisms satisfactorily evoked these qualities, which were increasingly seen as key to understanding the life of the human body. The clockwork universe had been replaced by a thermodynamic one, where the conversion of energy powered both the living body and the steam engine. Machines and living bodies continued to be understood as sharing a common nature, but key to that commonality was their 'self motive power', something both the cadaver and the clockwork automaton could not satisfactorily demonstrate. Anson Rabinbach's account of the importance of theories of energy to the understanding of modernity, *The Human Motor* (1990), notes the nineteenth-century disillusionment with eighteenth-century attempts to reproduce the dynamism of the human body:

> By the end of the eighteenth century, physiological vitalists like Paul-Josef Barthez delighted in denigrating the automata by pointing out that 'life' had eluded them and that their inventors had resorted to occult theories or 'not yet invented machines' to explain why they could not breathe 'life' into these artifices … With the invention of the steam and internal combustion engines, however, the analogy of the human or animal machine began to take on a modern countenance … [T]he eighteenth-century machine was a product of the Newtonian universe with its multiplicity of forces, disparate sources of motion, and reversible mechanism. By contrast, the nineteenth-century machine, modeled on the thermodynamic engine, was a 'motor,' the servant of a powerful nature conceived as a reservoir of motivating power. The machine was capable of work only when powered by some external source, whereas the motor was regulated by internal, dynamic principles, converting fuel into heat, and heat into mechanical work. The body, the steam engine, and the cosmos were thus connected by a single and unbroken chain of energy. (Rabinbach 1990: 51–2)

The anatomist's cadaver – with its limp, pliable limbs, emptied veins and stilled heart – was a far from perfect representative of the body as bearer of energy. But in the eighteenth century anatomy had been joined by a new discipline, which during the nineteenth century emerged from anatomy's shadow to be become highly influential (Braun 1992: 9). Where anatomy sought to map the body's topography, physiology sought to plot and measure its various processes, investigating the body as a "'bearer of forces and the seat of duration" … [,] a distinctly living and moving body … whose most telling characteristics elude static or death-oriented investigatory techniques such as autopsy or anatomical photography' (Cartwright 1995: 36).

Rather than poking around inside dead bodies, the physiologist sought to capture the processes of the body in action. For unfortunate animal test subjects, this often involved grisly interventions into the function of their living bodies, leaving the envelope of skin permanently open to permit observation of the processes taking place within, or the insertion of measuring instruments. Of an experiment in which Etienne-Jules Marey inserted a measuring device into the beating heart of a horse, Lisa Cartwright comments:

> This experiment marks a dual shift in methodology: a shift toward movement as a characteristic state of the body, and a shift toward graphic inscription as a means of recording interior processes. Significantly, these shifts simultaneously mark a move toward implanting a technology of observation directly into the body studied – a technique that joins technology and the living body rather than using technology to sacrifice the body for the sake of analysis. (Cartwright 1995: 24)

But some of these new ways of measuring and inspecting did not require invasive insults to the body's unity. While anatomy had produced a new way of understanding the body by opening it up to reveal previously hidden vistas, physiology deployed new technologies of observation and measurement that rendered even mundane perspectives on the body unfamiliar. Particularly notable in this regard is the physiologists' adoption of still and then moving photography as they became available over the course of the nineteenth century.[2]

Of course, there are numerous ways in which the camera can be employed to investigate the workings of the body – documenting the views made available by dissection or microscopy, for example – but of particular interest is the way in which it was able to transform an everyday view of the body – for example the sight of someone descending a flight of stairs – into something new and unfamiliar.

It could do this, of course, by allowing a shift in temporal scale. Technicians – most famously Eadweard Muybridge in the United States and Etienne-Jules Marey in France – could pull moments out of the flow of bodily processes and movements so that they could be retained and inspected at leisure, revealing details unavailable to the unaided eye. As a result of this treatment, the body becomes most importantly a generator of processes and movements, and new technologies – particularly photography – performed a kind of

---

2 The camera was, of course, only one of the machines utilised by physiologists. The development of physiology was fundamentally dependent upon the assembling of an arsenal of machines (Borell 1987), most of which do not have the familiarity of the camera. While the aesthetic dimension of photography makes it an effective illustration of the broader logic of physiological investigation, other crucial tools such as the kymograph also reflect the broader movement towards shifts in temporal scale and the capturing of fluctuations and movements beyond the reach of the human sensory apparatus (Borell 1987: 55).

Fig 3.1   Etienne-Jules Marey with George Demeny, *Untitled (Sprinter)*, 1890–1900. Gelatin silver print (15.4 x 37.2 cm). Gift of Paul F. Walter to MoMA. © The Museum of Modern Art/Scala, Florence.

chronological vivisection. Rather than breaking the unity of the body down into physical fragments like the anatomist, the new machines of physiology broke the body down into fragments of time, producing new ways of perceiving the body that were no less revolutionary than those produced by the Renaissance anatomy theatre.

Central to this new approach was the interaction of living body and machine. Where the automaton maker had sought to use machines to reproduce the movements of the human body, in the nineteenth century machines were applied to living bodies in order to measure and analyse them in a way impossible for the human perceptual apparatus. Rather than machines appropriating the familiar movements of living bodies, here machines rendered the movements of living bodies unfamiliar. The value of the machine lay in its capacity to capture the body in ways that broke the unities and flows registered by human perception into smaller constituent parts, and to do so in a way believed to be perfectly accurate and utterly objective and impartial. Lorraine Daston and Peter Galison have argued for a moral dimension to this mode of investigation:

> Mechanized science seems at first glance incompatible with moralized science, but in fact the two were closely related. While much is and has been made of those distinctive traits – emotional, intellectual, and moral – that distinguish humans from machines, it was a nineteenth-century commonplace that machines were paragons of certain human virtues. Chief among these virtues were those associated with work: patient, indefatigable, ever-alert machines would relieve human workers whose attention wandered, whose pace slackened, whose hand trembled. Scientists praised automatic recording devices and instruments in much the same terms … It was not simply that these devices saved the labor of human observers; they surpassed human observers in the laboring virtues: they produced not just more observations, but better observations. (Daston & Galison 1992: 84)

While not a physiologist himself, the eccentric photographer Eadweard Muybridge (born Edward Muggeridge) was a key figure in the development of the camera as an instrument of scientific research. He is most famous for having settled Leland Stanford's wager on whether all of a galloping horse's hooves ever leave the ground, and from 1872 Muybridge worked at the rail baron's private horse track on what would later become the grounds of Stanford University in order to catalogue the components of the horse's gait using high-speed photography. John Ott has already argued for the significance of the fact 'that the owner and manager of locomotives was interested in the mechanics of animal locomotion' (Ott 2005: 407); the 'iron horse' was already superseding the horse of flesh and blood, and the Industrial Age seemed increasingly capable of replacing living bodies with the bodies of machines. This sense of interchangeability was complemented by the use of photographic machinery to produce a mechanistic understanding of the movement of living bodies;

rather than creating an opposition between living body and machine, the new machinic way of visualising bodily movement reduced them to a shared set of mechanical principles.

> [T]he Muybridge photos encouraged viewers to imagine the horse, and by extension, all of nature, as merely another kind of machine. Fixed within a cold, monotonous grid of six, twelve, or twenty-four, the serial prints demanded that viewers reconceptualise a sweaty, snorting, quivering mass of horseflesh as a dynamo performing an endlessly repetitive sequence of actions. The text on the reverse of these prints featured a detailed, methodical, and technical frame-by-frame analysis of equine mechanics. The effect on zoopraxiscope-goers, obliged to experience the same 12 frames again and again, must have been even more pronounced. It was as if the mechanical means of gathering and reporting data had been transposed somehow onto the subject of study. (Ott 2005: 414)

In his important historicisation of vision in the book *Suspensions of Perception* (1999), Jonathan Crary contextualises this movement towards the mechanisation of vision against a nineteenth-century loss of belief in direct human visual access to the world. Where the optics of Descartes had cast vision as a passive reception of objective information (see Chapter 1), in the nineteenth century vision came to be understood as an active, constructive mode of perceiving. Technologies like photography allowed some of the work of vision to be done by the reliable and disinterested machine, rather than the fallible human eye.[3] It was even suggested that Muybridge's photographs were part of a process whereby the images produced by machines would 'gradually train the eye' to see more accurately – in other words, more like a machine (Ott 2005: 418).

It was Muybridge's work that impressed the value of high-speed photography on the trail-blazing physiologist Etienne-Jules Marey. To please his father, Marey had became a medical doctor rather than following his desire to become an engineer (Braun 1992: 2–3); however, he went on to approach the investigation of the body as an engineering problem, devising a succession of machines with the capacity to capture various kinds of data concerning the movements and processes of living things (Dagognet et al. 1992: 11).

Marey's most famous device was the 'chronophotographic rifle', which rapidly revolved a photographic plate in order to take twelve shots per second, and which had been inspired by the 'astronomical revolver' invented earlier by the astronomer Pierre Jules Janssen (Braun 1992: 55–64). Having begun with non-photographic methods of measuring physiological processes, Marey had been convinced of the value of photography by the work of Muybridge, but

---

3   Crary has elsewhere argued that a transformation of sight was brought about not only through machines expressly designed to produce new visual experiences; persistence of vision was identified partly through the experience of watching train wheels, cogs and other mechanised circular objects moving at great speed (Crary 1990: 112).

sought to exceed Muybridge in the rigour and accuracy of the photographic record he created (Kemp 2006: 305).[4]

The images captured by the photographic rifle and similar devices allowed the human eye to see the hidden components of which bodily movement was composed. In breaking the life of the body down into smaller systems of movement, rather than the static anatomical systems of the anatomists, Marey was clearly a mechanist – his book of 1873, *La Machine animale*, clearly recalls Descartes' *bêtes machines* and La Mettrie's *l'homme machine*. But he was a mechanist mindful of the limitations of the mechanist views that had come before him. In the words of François Dagognet, 'Marey always defended the theory of the "animal-machine," as long as this machine was no longer conceived as a simple assemblage of pulleys, wheels and wires, but rather as a veritable "animated motor," a living machine, at the source of activation (locomotion, voice and so on)' (Dagognet et al. 1992: 44).

Key to the mechanics of the body was its capacity for dynamism, its quality of originating its own energy and movement. While this perspective was one that clearly could not be satisfied within the investigative reach of dissection, however, it was more harmonious with the creation of automata; after all, the automaton maker sought to reproduce bodily movement and dynamism, however imperfect the results. In fact, Marey created automata himself: in 1868 he built a mechanical insect, and in following years workers at his laboratory created a series of mechanical insects and birds to demonstrate the soundness of the conclusions he had drawn from his studies on the mechanics of flight (Rabinbach 1990: 99).

Time and Motion

While quite different – perhaps even oppositional – to anatomisation in important ways, physiology and the use of machines to investigate the body as locus of movement and process nonetheless harmonise with its larger assumptions. The body is still understood as a kind of machine, but because machines are importantly characterised by a capacity to generate energy and movement, anatomisation is considered unsatisfactory as a method of investigating the machine body.

However, the metaphorical relationship between body and machine remains, and with industrialisation arrived a new breed of machines that impressed the observer with their industrious animation. While these machines had not been designed to look like living bodies, they had been designed to replace the movements of living bodies, making a parallel inevitable, and, furthermore, they were imbued with a capacity noticeably lacking in clockwork automata.

---

4   In return, it was Marey who suggested that Muybridge use the zoetrope to imbue his photographs with movement (Rabinbach 1990: 100–103).

Like the living body, they produced within themselves the energy that drove their movements. In the words of Christoph Asendorf, 'Unlike the eighteenth century, in which man became a machine, in the nineteenth, the machine is assigned human characteristics … The machine has become a subject, the individual its object …' (Asendorf 1993: 41).

The idea that the machine is a kind of body is inherently more troubling than the idea that the body is a kind of machine. The two beliefs are not independent from one another – in order to believe the former proposition, it is necessary to already believe the latter – but the former belief brings with it the fear that, rather than simply being able to simulate living bodies, machines might compete with or supersede them. This fear seemed well grounded during industrialisation; while the automata of the Enlightenment had existed in a realm outside everyday life and had possessed no productive capacity, industrial machines operated amongst the human population, working in close relationships with or replacing human workers. For Karl Marx, the industrial machine was an entity that replaced the worker by claiming his tools and wielding them in his place (Marx 1976: 496–7), while the mechanised factory was a kind of monstrous organism that absorbed human bodies into itself, 'a productive mechanism whose organs are human beings' (Marx 1976: 457).

The body of the human worker and the body of the machine were both understood to generate the energy that drove industry. Rabinbach draws attention to the appearance of a new common language for understanding the industrial machine and the labouring body:

> The human body and the industrial machine were both motors that converted energy into mechanical work … From the metaphor of the motor it followed that society might conserve, deploy, and expand the energies of the laboring body: harmonize the movements of the body with those of the industrial machine. (Rabinbach 1990: 1–2)

Although the automata that the likes of Descartes and La Mettrie considered so instructive regarding the life of the body were no longer so compelling, a whole new ecosystem of machines had arrived, one which further weakened the division between biology and mechanism.

With human bodies seen as machines, their actions and energy increasingly interchangeable with those of fabricated machines and productively inserted into larger mechanical/biological systems, the practical value of applying the kind of scientific investigations exemplified by Muybridge and Marey to labour became apparent. From 1894, research took place at the Institut Marey (under Marey and then his successor as director Auguste Chauveau) that sought to apply techniques for the capture and quantification of movement to the optimisation of the labouring body – initially that of the blacksmith (Braun 1992: 324). Twenty years later, Frank B. Gilbreth would present a series of photographic studies of labour that served a similar purpose, and this work was part of the broader twentieth-century rise of the time-and-motion study (Rabinbach 1990: 117, Cartwright 1995: 37–8), an effort to create a 'science' of work initially championed by both Frederick Taylor in the United States and comparable approaches in Europe, which

> took as their starting point the decomposition of each task into a series of abstract, mathematically precise relations, calculable in terms of fatigue, time, motion, units of work, and so forth … [,] economizing motion and achieving greater work performance through adapting the body to technology … And above all, each shared the Utopian hope that it was possible to resolve industrial conflict scientifically and rationally in the interests of economic progress. For both it was the body of the worker that constituted the point of contestation in the industrial sphere, not the politics of workers' organizations or ideological issues. (Rabinbach 1990: 242–3)

The supposed effectiveness of Gilbreth's photographic approach to this project again hinged on the use of machines to mechanistically render the body. Where Taylor's earlier time-and-motion work had relied on stopwatches to quantify the efficiency of workers, and had consequently been attacked on the grounds that human stopwatch operators were unreliable, Gilbreth employed cameras, machines whose inhuman gaze 'ostensibly captured everything and missed nothing, "trapping" the motions of their subjects in what was apparently the unambiguously objective frame of motion study' (Gainty 2012: 5). As with Stanford's racehorses, photographic machinery produced representations that transferred its mechanistic properties to its living subjects.

The attempt to re-engineer the worker's body so as to insert it into a more seamless and productive relationship with machines is exemplified by the rise of the production line. In his memoir *My Life and Work*, Henry Ford – perhaps the patron saint of the production line – holds himself up as greatly concerned with the wellbeing of his workers, but his concern seemingly revolves around their right to a kind of improving, even symbiotic, relationship with the machines in his factories: 'I have been told by parlour experts that repetitive labour is soul- as well as body-destroying,' he states, 'But that has not been the result of our investigations' (Ford 1922: 105).

> I could not possibly do the same thing day in and day out, but to others, perhaps I might say to the majority of minds, repetitive operations hold no terrors. In fact, to some types of mind thought is absolutely appalling. To them the ideal job is one where the creative instinct need not be expressed … The average worker, I am sorry to say, wants a job in which he does not have to put forth much physical exertion – above all, he wants a job in which he does not have to think. (Ford 1922: 103)

Ford imagines the rhythms of his machines soothing the minds of his witless employees with their predictable monotony of movement. While the prospect of such work horrifies Ford himself, he imagines that 'the average worker' finds it a relief to enter into a relationship with a machine that frees him from the obligation to act and think independently. But Ford found that he could take this symbiotic relationship further still because his machines could function not only without the guidance of a complete brain. Because of the restrictive way in which they engaged the bodies of his workers, they did not even need a complete body:

> It turned out at the time of the inquiry that there were then 7,882 different jobs in the factory. Of these, 949 were classified as heavy work requiring strong, able-bodied, and practically physically perfect men; 3,338 required men of ordinary physical development and strength. The remaining 3,595 jobs were disclosed as requiring no physical exertion and could be performed by the slightest, weakest sort of men. In fact, most of them could be satisfactorily filled by women or older children. The lightest jobs were again classified to discover how many of them required the use of full faculties, and we found that 670 could be filled by legless men, 2,637 by one-legged men, 2 by armless men, 715 by one-armed men, and 10 by blind men. (Ford 1922: 108)

The Ford plant might reduce a worker to nothing more than an arm endlessly pulling down a drill press, so what matter if the worker had a second arm, or any legs at all? With the efficient incorporation of human bodies into the larger organism of the car factory, a great deal of the substance of those bodies became simply redundant, a kind of optional extra that was of no particular value. Good news for a crippled man otherwise incapable of employment during the Great Depression, perhaps, but also indicative of the way in which mechanisation refigures the role of the labouring body, making it one more engine driving small, repetitive movements within a larger mechanism (see Seltzer 1992: 157).

It is therefore not simply the creation of more complex or autonomous machines that leads to a stronger sense of commonality between body and machine in the Industrial Age; it is also the fragmentation and limiting of the bodies of workers. While La Mettrie's promised speaking machine might have remained unattainable, Ford's production line worker did not need to speak anyway, and did not even require the level of manual dexterity displayed by Vaucanson's flautist. The figure of the robot appears not simply in a milieu of more sophisticated machines (after all, a convincing mechanical simulation of the human body was as impossible in the early twentieth century as it had been in the late eighteenth), but one of human–machine convergence. In certain settings, at least, an increase in the power, complexity and autonomy of machines was accompanied by a complementary decrease of these attributes in the humans around them. The (metaphorically or literally) one-armed worker of

Ford's production line prefigures those one-armed machines like the Unimate that will one day replace him.

Machine Men

However, it would be wrong to conclude from this that the robot necessarily originates in a devaluing of the human body by modern technology and labour practices. Take, for example, the following quote regarding the position of human workers in globalised technology production:

> The human body, because it is placed outside the rationale of value, is seen as something whose value is so incommensurable and therefore immeasurable that it ultimately does not cost anything and so is socially devalued. As nonvalue, it therefore always costs less than technology (this is the reason for the slowing down of robotics). (Fortunati et al. 2003: 3)

This incommensurability is made abundantly clear by the relative economic worth ascribed to computers and the developing world bodies that assemble them; however, the parenthetical remark concerning robotics requires clarification. For 'robotics' perhaps it would be more accurate to substitute research into androids, that subset of robots that reproduce the human form; this would seem to be the spirit of the quote given the parallel with the human body, and the creation of robots like the robot arms of the car assembly line or microprocessor factory, which do not aim to simulate the entire human body, is a highly successful and economically important industry. It is the attempt to create a robot that fully reproduces the attributes of a whole human body that has for many years failed to deliver something approximating the fantasy that motivates it.

However, in this regard I would argue that Fortunati et al. are failing to appreciate an important dimension of attempts to create androids. It is true that there is little economic imperative to create mechanical replacements for human bodies – after all, the world is full of real human bodies, which do not need to be designed and built at great cost – but this only raises the question of why such research has taken place at all, rather than requiring explanations for its lack of success. The quest to build a machine that could walk began before the invention of the steam train (Asendorf 1993: 105), and yet Honda's ASIMO ('Advanced Step in Innovative MObility'),[5] the first bipedal android capable of reliably nimble perambulation, only attained this capacity in 2001 (Honda 2006). The slow progress towards ASIMO resulted not from a lack of economic imperative, but rather from the fact that reproducing the countless adjustments of balance necessary for human beings to teeter about on two long legs for

---

5   This rather tortured acronym is obviously an attempt to create a reference to Isaac Asimov, the famed science fiction author associated with robot-themed speculation.

many years seemed an insurmountable engineering challenge.[6] Again, this invites the question of why this problem was pursued for so long when robots can just as easily be fitted with wheels, or a far less challenging four (or more) legs. Rather than reflecting a devaluing of the human body, these efforts suggest an idolisation, a belief in the perfection of the human form, which gives rise to a demand that machines conform to its features. Furthermore, at the time that they were writing, Fortunati et al. were presumably unaware that a resurgence of robotics, which had begun in the 1990s, was gradually gaining momentum; after decades of false starts and dead ends, a viable android industry was beginning to look like a realistic possibility for the not-too-distant future.

While the quest to build androids might be cast as driven by the economic needs of particular historical moments, Fortunati et al.'s comment only draws attention to the weakness of these justifications. While robotics has been pursued most vigorously in Japan, a country whose past rapid economic growth and resistance to immigration can be seen to have created an economic imperative for new sources of labour, there is a great gulf separating the robotic arm on a car assembly line from ASIMO. The fact that the first wave of robots that were lifelike enough to capture the public's imagination – creations such as Sony's AIBo ('Artificial Intelligence roBOt') animals (Sony 2005) and Honda's ASIMO android – were automata with no real productive use only serves to accentuate this fact. The AIBo's claimed *raison d'être* was its usefulness as a companion, but the argument about the incommensurability between the value of the human body and its robotic simulation is surely doubly true when the human body is substituted with that of a dog or cat, and production of AIBos was ended by Sony in 2005 because of their economic non-viability (Anon. 2008b). Also cancelled were Sony's QRIO androids, which seemed to have been productive only as tools for studying human–robot interaction; in other words, facilitating the study of how humans might feel about living in a world filled with useful robots rather than demonstrating any usefulness that might justify the motivating assumption that this world will one day arrive.

It has been argued that androids have a superior capacity to function in the home, helping to care for Japan's increasing proportion of elderly citizens – a human-like robot would be best suited to operating in a domestic environment designed for human use. However, the ASIMO project was initiated in 1986, a time when Japanese technological research was driven more by bubble economy hubris than concern about physical frailty, and the economic impact of a rising average age was only just beginning to be recognised.[7] During ASIMO's

---

6   And, in fact, ASIMO still does the job poorly and inefficiently compared to a human body (see Clark 2008: 3–11).

7   For example, a Japanese report on population aging's economic impact produced for the United Nations in 2004 cited no research in this area earlier than 1987 (Shimasawa & Hosoyama 2004: 4).

various public appearances, its capacities are shown off through economically insignificant activities such as dancing, rather than opening jars for arthritic old ladies or carrying them upstairs to bed.[8] At the 2005 International Robot Show in Tokyo, 80-year-old Joe Engelberger, one of the founders of the world's first industrial robot manufacturer in the 1950s, was reported to have complained about the state of robot research, which he characterised as preoccupied with making 'toys' or 'dolls' that mimic human bodies rather than producing machines capable of practical application (Cameron 2005: 21). The fact that both AIBo and ASIMO have been used primarily in corporate advertising for their respective producers, demonstrating their makers' technological resources and know-how, rather than being seriously marketed as consumer goods, further makes the suggestion of an economic imperative in their development seem no more persuasive than claims for the aerodynamic properties of fins on 1950s cars. This is not to argue that such robots can have no productive function, but only to note that their use value seems largely unrelated to their appeal. Just as 1950s industry can be seen to have made reference to technological fantasies of streamlined speed with its fins, the android can be seen as resulting more from a deep-seated symbolic association than any pragmatic justification attached to it after the fact.

ASIMO illustrates this point well because its popularity rests on the tendency of its audience to misinterpret what they are seeing during its public appearances. ASIMO seems to be acting more-or-less independently as it interacts with its environment or human minders, but it is, in fact, remotely controlled by a human operator (Brooks 2002: 71).[9] Its shape and mode of movement lead naïve observers to credit it with a human-like agency and understanding that it lacks (and it is intentionally presented in a way calculated to encourage this misattribution), but ultimately it is only able to simulate human bipedal gait, rather than any other human capacity. ASIMO impresses its audience partly because of what it is, but also partly because its audience doesn't understand what it is.

The expectations of and promises made by robotics are often dependent upon a lack of clarity concerning what precisely a robot is the *Oxford English Dictionary* defines a robot as both:

1a. An intelligent artificial being typically made of metal and resembling in some way a human or other animal...[10]

and

---

8   A collection of images showing ASIMO in action can be seen at http://asimo. honda.com/gallery/.

9   The latest ASIMO model now has some capacity to carry out pre-programmed instructions, but is still primarily controlled by an operator using a laptop computer.

10   Leaving aside definition 1b, which has already been discussed above.

2. A machine capable of automatically carrying out a complex series of movements, *esp.* one which is programmable.

It is the confusion of those two definitions that leads to much of the difficulty. In the popular imagination, nourished on science fiction robots, the first definition is key. For specialists, the second definition is most important, and being like a human – while it might be considered desirable – is not considered definitional. But in reality the two definitions are constantly confused with one another.[11]

Various machines are referred to as robots despite not being remotely human-like: the Roomba robot vacuum cleaner is clearly not an android (although it does seek to replace human labour), but devices like the Roomba can carry out relatively complex actions automatically, independent of human control, and this is taken as the basis for its qualifying as a robot. However, if we adhere solely to this criterion, things become more and more confused as the technology in our lives becomes ever-more autonomous. For example, I don't suppose anyone would call the anti-lock braking system in my car a robot, and yet it reacts to its changing environment with a high degree of autonomy; it could save my life without my even realising that it has done so. Should our definition of robots expand over time, therefore, until almost all the technology in our lives qualifies for the title?

In the popular imagination, this just won't do. Take another example of an automated machine. At the beginning of the 1960s, Joe Engelberger introduced the Unimate, the first robot arm, into car manufacturing, and these robot arms have since become commonplace. Almost everybody has seen one, and everyone agrees that they are robots. But such devices actually have little autonomy – they are really just jointed metal lifters worked by motors and controlled by computers, performing the same repetitive tasks over and over again, incapable of reacting to anything in their environment. Compare the Unimate with a laser-cutting machine. The laser-cutting machine can cut various materials with a high degree of precision, using a motorised frame to move a high-powered laser under the control of a computer. It has the same degree of independence and control as the robot arm, and so, going on the second definition of a robot, has an equal claim to being one. But nobody calls laser-cutting machines robots. Why? The answer is obvious: the action of the laser-cutting machine is not evocative of the human body. The robot arm in a car manufacturing plant isn't even that close a recreation of the human arm, but it moves in a way that invites analogies with a human arm, while the laser gliding over its frame in the

---

11   It is particularly striking that the second meaning, which clearly arises from an overly-generous interpretation of the first sense, was already in use before the end of the 1920s, the decade in which Karel Čapek introduced the term 'robot' (see below), demonstrating how eagerly the concept was taken up and expanded.

laser-cutting machine invites no such analogies. Returning to ASIMO, it only partially satisfies the first definition of robot (not being intelligent), but also, being remotely controlled by a human operator, satisfies the second definition partially at best – certainly less convincingly than the laser cutter. Our belief that it qualifies as a robot is almost entirely dependent upon the fact that it looks and acts like a human body.

So the two definitions often co-exist, and it is easy to slip between them. People building or marketing machines are often tempted to blur the distinction, leveraging the public's fascination with the idea of artificial people to generate interest in autonomous machines, and over-optimistic predictions regarding robots often sustain themselves precisely by fudging the difference between these two senses of the term, using the existence of highly autonomous machines (definition 2) to support claims about the viability of intelligent human-like machines (definition 1). While a given robot might satisfy both definitions, it is misleading to use them interchangeably.

Ultimately, the application of the term 'robot' to a machine seems to reflect an affective, rather than technical, criterion. We call something a robot if it evokes some kind of identification or fellow-feeling; if something about it – its animation, agency, or simple shape – triggers an association with the human body. Of course, this fellow-feeling often fades over time; clockwork automata lost the capacity to inspire it, and one of the first applications of the term 'robot' to actual machines in English was to the automated traffic light – a device with little capacity to amaze us with its seeming autonomy today. The more we watch such machines, the more aware we become of their meager stock of actions and reactions, and so the less agency and autonomy they seem to have. Still, to see something as a robot, we must – even if only for a moment – fall victim to the illusion that we are seeing something meaningfully like a living body.

Robot Lovers

The term robot itself was introduced in 1920 by Czech writer Karel Čapek in his play *R.U.R.* (*Rossum's Universal Robots*). Čapek's play – presumably inspired to at least some degree by the then-recent revolution in Russia – depicted a population of artificially created beings designed for servitude who, when later imbued with the capacity for emotions and independence, rise up to destroy humanity. Čapek imagined his robots as engineered biological beings created from 'a substance which behaved exactly like living matter, although its chemical composition was different' (Čapek & Čapek 1961: 5) rather than machines (more the replicants of *Blade Runner*, perhaps, than the killing machines of the *Terminator* films), but the story clearly connected with popular anxieties concerning the increasing mechanisation of industrialised societies, and productions of the play depicted the robots as humanoid machines regardless.

The play was translated, and performed in New York in 1922 and Tokyo in 1924 (Moran 2007b: 1399, Schodt 1990), and the robot was soon established as a human-like machine, and the story as a template for tales of its dangers for humanity. In 1935 Čapek railed against this reinterpretation of his story, stating, 'I recoil in horror from any idea that metal contraptions could ever replace human beings and awaken something like life, love, and rebellion. Such a grim outlook is nothing but an oversimplification of the power of machines and a grave insult to life' (cited Moran 2007b: 1401). But despite Čapek's protestations, the robot was fixed in the popular imagination as a machine made to serve humanity but which threatened to usurp its masters. This new vision of the android was tied to both the increased power and sophistication of machines, and the increasing degree to which they either replaced human labour, or were incorporated into human labour in a way that required human beings to behave in ways designed to complement them. And yet there is clearly as much desire as fear in accounts of the robot. Robots are not presented as the greatest threat in such science fiction stories – rather it is the human desire to create them. Robot disaster narratives suggest that the robot is a known threat, its eventual rebellion expected from the start, and the responsibility for the rebellion ultimately lies with the human creator who unleashes the threat regardless. In other words, it is an uncontrollable human compulsion to create mechanical bodies that poses the greatest danger. Time and again, both positive and negative accounts of the robot, both those in fiction and those in allegedly plausible predictions, suggest that the creation of human-like machines cannot be avoided – for good or ill, it is humanity's inescapable destiny.

David Levy's 2007 book *Love + Sex with Robots: The Evolution of Human-Robot Relationships* takes the uncritical fantasies we often entertain about fabricated bodies to absurd lengths. As a result, it is useful as an example of the ways in which a combination of wishful thinking, false logic, and fuzzy definitions is routinely employed to satisfy a longing for the existence of human machines. The book is as strange as its title would suggest, filled with leaps and gaps of logic and inexplicable assumptions, but its subject matter – a combination of the salacious and the weird – guaranteed it a certain level of notoriety. It opens with a vague account of the history of 'robotics', which, as tends to be the case with such accounts, seeks to establish a robot future as the inescapable endpoint of a millennia-old human enterprise. From Greek Alexandria onwards, it is suggested that any attempt to make anything that looked like a living thing and moved was part of a unified quest to produce artificial life, a quest now nearing completion. As already noted, in reality the motivations and understanding behind history's assorted efforts of this kind have been extremely varied, and up until the eighteenth century – at least – the very idea of creating a thinking, artificial mechanical person would have most likely been incomprehensible even to those involved in them.

The idea that the creation of androids is a future inevitability is a common assumption, but Levy takes this fantasy further by claiming that this future has already somehow arrived unnoticed. This belief is facilitated by the vagueness of terms such as 'robot' and 'artificial intelligence', and a general tendency to apply them to any example of technology even remotely reminiscent of the science fiction ideal. Most importantly, the two alternative definitions of the word 'robot' – an intelligent artificial being and a machine capable of complex actions – are conflated to make automated movement synonymous with human-like intelligence and even emotional capacity.

Levy uses the term 'robot' to refer to such things as electronic toys and dolls, automated vacuum cleaners and lawn mowers, university robotics experiments, industrial machines, wearable power-assisted suits, and mobile information kiosks, and makes no differentiation between such things and the thinking artificial human beings both he and his readers understand to be at issue, creating the impression that the self-aware robots of science fiction have already been created. Clearly what is most at issue here is the possibility of creating machines that can think: Levy (like many other enthusiasts) treats the future arrival of machines in possession of something like human consciousness as an inevitability, despite the fact that attempts to create human-like robots to date do not at all establish this. The key point in relation to Levy's account, however, is the suggestion, again, that artificial intelligence has already been successfully created, at least on a modest scale.

Since the birth of the science of artificial intelligence in the mid-1950s, gigantic strides have been made in the quest for a truly intelligent artificial entity. The defeat of the world's best chess player, Garry Kasparov, was just one of these strides. Others include the creation of computer programs that can compose music that sounds like Mozart or Chopin or Scott Joplin, at the operator's behest; programs that can draw and paint better than many human artists whose work today hangs in art galleries and in the homes of wealthy collectors; and programs that can trawl the Internet and write news stories based on the information they gather, stories written in a style of which most journalists would be proud. Then there are expert systems – programs that incorporate human expertise to enable them to solve analytical problems normally assigned to human experts. Such programs are powerful tools for medical diagnosis, and they have also proved to be highly competent in a wide diversity of other fields, such as prospecting for minerals, making political judgments, detecting fraudulent uses of credit cards, and making recommendations in court cases to judges and lawyers, even advising defendants how to plead. These are not examples of what might be in the future – they are just some of the accomplishments of AI in its first fifty years. (Levy 2007: 6–7)

Again, this treatment avoids definitions, detail or differentiation in order to take advantage of a lack of clarity in what precisely is being discussed. 'AI' today is attributed to all manner of sophisticated computer programs, including the computer games on which Levy is an expert, but these computer programs are clearly something qualitatively different from artificial intelligence as it is both popularly understood and as Levy discusses it in relation to his imagined robot companions of the future.[12] More than this, he slips once more into fantasy with his claim that computers have already made artists, musical composers and journalists redundant through their ability to create art and music and write prose with the same (or better) level of skill as human beings.[13]

Japan serves a valuable role for Levy in his fantasies concerning the robotopia of the present. Japan is of inevitable value in such a discussion because of both the amount of research into robots being conducted there and the stubborn adherence to robot-related fantasies that motivates it. However, for Levy, Japan serves as a robot fantasyland; it is presumably where all the mechanical painters, musicians and journalists he doesn't run into in his own neighbourhood live:

> At first, service robots were mainly used for drudgery-related tasks – cleaning robots, sewer robots, demolition robots, mail-cart robots, and robots for a host of other tasks, such as firefighting, refueling cars at gas stations, and in agriculture. But after the service-robot industry became well established in Japan, the country's robot scientists turned their attentions to the realm of personal robots, to be used at home by the individual. Mowing the lawn and vacuuming the carpet have both become tasks that in a slowly but steadily increasing number of homes are now undertaken by robots. Similarly, robots are beginning to be used in education, and Toyota has announced that by 2010 the company plans to start selling robots that can help to look after the elderly and to serve tea to guests in the home. (Levy 2007: 7)

This passage is also an illustration of Levy's tendency to turn speculation about the future into a history lecture ('But after the service-robot industry became well established in Japan …'). Historical events that have yet to happen are described as features of the past from which we can make extrapolations of the future. Levy even builds upon this technique by transforming speculation from earlier in the book into historical fact later: in the above excerpt he uses the

---

12 Artificial Intelligence will be discussed in more detail in the following chapter.

13 In this Levy is repeating claims made previously by Ray Kurzweil (who will be discussed in more detail in the following chapter). Of course, it is meaningless to talk about a computer artist in this sense, as the computer has no intention and is aiming for no particular effect. The program is simply created by a human being to simulate a practice that other human beings might find aesthetically satisfying. To claim that this program is an artist is like claiming that the microbes that make coral are artists, given that they, too, are capable of producing artefacts human beings find aesthetically pleasing.

supposed historical reality of robotic helpers working in the service industries to support the claim that soon robots will live in our homes and help the elderly; shortly afterwards he argues that 'just as robots have evolved from assembly-line machines to companions for the elderly, so pets have also evolved into our companions' (Levy 2007: 46); having introduced the idea that robots might become helpers for the aged, he can then treat this phenomenon as a matter of historical record analogous to the human domestication of animals in prehistory.

Once again, by counting any automated mechanical system as a robot and importing it into a discussion of conscious androids, Levy depicts Japan as a land where robot firemen appear to battle conflagrations, robot petrol station attendants fill your car, and a robot gardener takes care of your lawn. In reality, for all the plans and predictions of Japanese companies, the reality is disappointing in comparison.

This foundational assumption that human-like robots are already a reality, whose arrival on a flight from Japan can be expected imminently, explains the rather idiosyncratic angle of this book on robots: while other commentators might debate the possibility of one day creating robots that possess a level of independence and physical prowess necessary to replace menial human labour, Levy seems to think that the robots of science fiction novels have already been successfully created, and so the only remaining question is to what extent they can be incorporated into our physical and emotional lives.

Levy's answer is, predictably, that they will be incorporated into our physical and emotional lives to the same extent as human beings. To support this claim he presents a lengthy and meandering discussion of some examples of research into causes of love whose relevance to the issue at hand is frequently unclear. The justification he does put forward for his position does not progress much beyond his initial argument that: love is a form of attachment; people can become attached to objects such as computers; therefore, people can love computers (and, by extension, robots, although it isn't explained why people don't fall in love with the computers they already have in their houses) (Levy 2007: 25–30).

Having presented this rather flimsy argument, Levy then proceeds – without segue or justification – to operate on the assumption that, not only *can* human beings fall in love with robots, but robots will be built in such a way as to *make* human beings fall in love with them:

> The uniqueness of the 'in-love' brain scans could serve as the basis for robots
> to determine whether or not a particular human was falling in love with them. A
> robot who wants to engender feelings of love from its human might try all sorts
> of different strategies in an attempt to achieve this goal, such as suggesting a visit
> to the ballet, cooking the human's favorite food, or making flattering comments
> about the human's new haircut, then measuring the effect of each strategy by
> conducting an fMRI scan of the human's brain. When the scan shows a higher

measure of love from the human, the robot would know that it had hit upon a successful strategy. When the scan corresponds to a low level of love, the robot would change strategies. (Levy 2007: 36–7)

It's a dramatic shift from automatic vacuum cleaners to android sociopaths intent on manipulating our emotions, who periodically ask us to lie down inside the minivan-sized fMRI machines unobtrusively incorporated into their bodies so that they can evaluate their success. But Levy makes that shift without any intervening discussion or evidence.

How does Levy know that such robot behaviour will be possible? Well, it will be 'programmed' into the robot, of course. He states that, 'Not only will the robot be programmed and learn to have similar interests and other characteristics as its human owner, it can also be guaranteed by its programming to find its owner emotionally attractive' (Levy 2007: 44). For someone presumably very familiar with computer technology, Levy's assumptions are surprisingly reminiscent of accounts of computers from the 1960s or 1970s, when these disturbing alien devices were credited with the capacity to do almost anything as long as the right punchcard was fed into them.

How might one program a machine to have personal interests of any kind, let alone ones matching a particular person, or guarantee that it will find someone 'emotionally attractive'? Levy can have no more idea than anyone else but seems confident that, using the magic of 'programming', our future robot companions can be made to think and feel whatever we wish with little effort. For Levy, the outcome of all this is beyond doubt:

> [M]y thesis is this: Robots will be hugely attractive to humans as companions because of their many talents, senses, and capabilities. They will have the capacity to fall in love with humans and to make themselves romantically attractive and sexually desirable to humans. Robots will transform human notions of love and sexuality. I am not suggesting that most people will eschew love and sex with humans in favor of relationships with robots, though some undoubtedly will. But what *does* seem to me to be entirely reasonable and extremely likely – nay, inevitable – is that many humans will expand their horizons of love and sex, learning, experimenting, and enjoying new forms of relationship that will be made possible, pleasurable, and satisfying through the development of highly sophisticated humanoid robots … Humans will fall in love with robots, humans will marry robots, and humans will have sex with robots, all as (what will be regarded as) 'normal' extensions of our feelings of love and sexual desire for other humans. Love with robots will be as normal as love with other humans, while the number of sexual acts and lovemaking positions commonly practiced between humans will be extended, as robots teach more than is in all of the world's published sex manuals combined. Love and sex with robots on a grand scale are inevitable. (Levy 2007: 22)

97

Levy is clearly confident in his conclusions. In fact, he feels all this to be so self-evidently true and based upon such sound scientific reasoning that he likens anyone with the temerity to express scepticism regarding his claims to those who reject the theory of evolution (Levy 2007: 20). Meanwhile, back in Japan, the country where Levy's robot revolution supposedly has already happened, how are things coming along?

Well, in 1998, the powerful Japanese Ministry of Economy, Trade and Industry (METI) unveiled an ambitious five-year, five billion yen research initiative dubbed the Humanoid Robotics Project, which brought together the resources of 'The National Institute of Advanced Industrial Science and Technology, [The] Manufacturing Science and Technology Center, 11 university laboratories and 12 corporations' (Kusuda 2002: 415). Its goal was 'to develop first-generation intelligent humanoid robots, able to use hand tools and work in human environments, including hospitals, offices and households' (Robertson 2010: 11). An account of the state of this project written in the fourth of its five years might have served as a humorous epitaph if not for the fact that the project refuses to die:

> A symposium was held in February 2002 to make a[n] interim report on the current status of the ongoing project. At the end of the symposium a panel discussion was held with panelists … Professor Inoue, [of] the University of Tokyo, as the chairman of the project[,] and participating researchers from the National Institute of Advanced Science and Technology, Honda and Sony. An audience [member] asked 'In 2005 will a humanoid robot be working in my home?' The answer was [a] definite 'No'. Then, 'When will [h]umanoid robots be in use in my house?'[,] 'In which area will humanoid robots find applications?[']
>
> The answer was again very low key. Professor Inoue gave '2025 or 2030 I don't know'[,] 'I cannot name the possible applications. Entertainment area will be a possibility'. The chairman of the panel discussion was frustrated enough to request to the researchers[,] 'At the finalizing report of this project next year please be prepared to give us some indication on when and how humanoid robots will be used in the future'. (Kusuda 2002: 416)

In the same year, *Time* magazine reported: 'The Japanese government's Humanoid Robotics Project set out five years ago to deliver a robot versatile enough to perform hard labor in hazardous conditions. Some $40 million has been spent but the project's HRP-1 robot still suffers from poor visual recognition and has trouble walking on rough terrain' (Lee 2002).

The conclusion to draw from the HRP project is not that robots do not live among us because human bodies are a less valued resource, as suggested by Fortunati et al. but rather that many people – including those in powerful corporations and government ministries – remain desperate to create a world in which robots *do* live among us, despite their history of investing tremendous amounts of money and time in exchange for generally demoralising results. For the record, the Humanoid Robotics Project continues today, major Japanese corporations such as Honda, Sony and Matsushita continuing to invest resources more than ten years into their five-year plan. On a visit to the programme for the 2008 BBC *Horizon* documentary 'Where's my Robot?' the then-latest product of the Human Robotics Project, HRP-3, was put on show, but the demonstration was ended abruptly when the robot fell over on a perfectly flat interior floor and could not get up, causing staff to hide it behind a makeshift screen while the interviewer and crew were escorted from the building (Anon. 2008a).

But HRP androids keep coming. The latest model, HRP4c, represents a marked shift in approach and priorities for the project. Rather than working in hospitals and disaster zones, the HRP4c, which has been given a feminine shape and rubber female face, has been demonstrated modelling clothing on a catwalk and dancing jerkily to pop music. Rather than abandon the project in light of its costly failure to deliver any of its intended outcomes, it would seem that researchers instead have sought to create a robot that can attract media attention through Youtube clips in spite of its vastly inferior capabilities.[14]

Bodies for Machines

The HRP4c, whose feminine body symbolises an abandonment of attempts to create a fully-fledged artificial person in favour of decorative entertainment, can be placed within a larger history of making machines appear more 'human' and non-threatening. Through her research into Japanese robots and roboticists, Jennifer Robertson has concluded that a desire to imbue robots with gender is reflected in the relatively greater 'streamlining' of the feminised robot's body. Although this streamlining is also apparent more generally in the creation of appealing robots, the 'attribution of female gender requires an interiorized, slender body, and male gender an exteriorized, stocky body' (Robertson 2010: 19).

Iconic machines of the nineteenth century such as steam engines were admired for the inhumanity of their rational designs, but by the time of Čapek's play and Ford's production line the incorporation of machines into daily life had eroded any sense of separation between body and machine. It was not long before the design of technology began to make quite direct references to organic shapes and contours. The earlier aesthetic appreciation of the non-organic, engineered, powerful look of the big machine, and the form-follows-function and engineering-as-art sensibilities that informed movements like Functionalism, the Bauhaus, and Futurism in their different ways, later came to be replaced by a smooth, sleek exterior. In Stuart Ewen's words, where

---

14   See, for example, http://www.youtube.com/watch?v=xcZJqiUrbnI.

'[m]odernism had once called for a unity of form and substance, now it had become a fluidly suggestive shell, wrapped around an inner mechanism that was hidden and mystified' (Ewen 1988: 148). Accounts such as Ewen's note the movement away from the uncompromising look of unadorned functionalist engineering towards a design that made machines look less like machines, and so less alien and threatening. Where the icon of the late nineteenth century had been the Eiffel Tower's bouquet of iron girders, in the 1920s the Empire State Building's advanced engineering was clad in a sleek Art Deco exterior. To quote Ewen again, 'With streamlining, the machine was invested with a soul. Its surfaces were decidedly metallic, but its forms were seamless and rounded, organic ... [T]his was a vision of a machine that was more spherical, softened, humanized; purged of its mechanical complexities and its threatening angularity' (Ewen 1988: 148). However, the idea of a machine soul seems an odd one; in the realm of design, with its concern for shape and surface, there is little room for the immaterial. The unification of objects through an aesthetic of streamlining – the containment of machinery within an exterior skin – is more properly about the body, rather than the soul. This referencing of the body makes the machine's evocation of life seem less threatening and alien by making it appear closer to the familiar life of human bodies.

While it seems clear that streamlining and related design aesthetics make quite explicit reference to the body, I nevertheless don't find the account of Ewen and others entirely satisfactory, as they don't seem to pursue their discussion of the cultural production of aesthetics far enough. While understanding the appearance of machines as – of course – dictated to a significant degree by culturally produced aesthetic sensibilities, such accounts say nothing about the body referenced by such sensibilities. In other words, while the appearance of the machine is seen to reflect historically and culturally specific factors, the 'seamless', 'soft', 'rounded' unified living body being referenced by this design style is presented without comment, seemingly as something self-evident and natural.

Jean Baudrillard similarly evokes a 'natural' unity and smoothness that industrial design seeks to appropriate for machines, which are understood as otherwise characterised by opposing characteristics: 'As in the development of some animal species, the form is externalized, enclosing the object in a sort of carapace. Fluid, transitive, enveloping, it unifies appearances by transcending the alarming discontinuity of the various mechanisms involved and replacing it with a coherent whole' (Baudrillard 1996: 53–4). But the bodies being referenced in this way are no less the product of culturally produced aesthetic sensibilities than the machines that reference them. While streamlining seeks to hide the alarming, fragmentary complexity that animates the machine, the human body is understood to be animated by inner workings that, if anything, are *more* fragmentary and complex than those of the machines around us, and certainly cause more alarm when exposed to view. In fact, the earlier connections made between human body and machine by the mechanical philosophy of Descartes' day hinged on precisely this similarity: investigating the human body through dissection, which laid bare its inner workings, the mechanical philosophers saw these inner systems as functioning like the gears, pumps and pulleys that animated machines.

Rather than taking a machine that seems fundamentally unlike the body and attempting to make it like a body, then, streamlining might be better understood as motivated by a foundational association of machine *with* body, which then engenders attempts to make that artificial body reflect the same aesthetic values to which living bodies are expected to conform. The rise of the streamlining aesthetic in the twentieth century was accompanied by phenomena such as Frederick Taylor's time-and-motion studies and Henry Ford's production lines, which sought to instil rationality, efficiency, and reliable repetition into the bodies of workers, even as machines – previously marked by their austere utilitarianism – were sealed within flowing, moulded 'skins' and even decidedly non-utilitarian nose cones, flanges, or fins.

Therefore, as the bodies on Ford's production line or in Taylor's time-and-motion studies were being broken down into discontinuous fragments by the demands of industrial production, the internal mechanisms of machines were being gathered together, unified, and enclosed in outer skins. There is no reason to see these movements as taking bodies and machines in opposing directions, however; the neo-classical nude gathered the jumble of mechanisms found within the body by anatomy and sealed – but did not hide – them within a beautiful unifying exteriority. These two movements are complementary, deepening the reciprocity between machine and body; as the body is understood to become more like a machine on the shop floor, the machine is understood to become more like a body in the home. The body becomes more utilitarian and defined by energy and repetition in industry, but the machine becomes more aestheticised and independent of concerns regarding utilitarianism and power as commodity culture makes it the subject of emotional investment and aesthetic discernment.

Even as machines were being imbued with qualities associated with human bodies, there were attempts to imbue human bodies with qualities associated with machines, and both movements were inflected by historically specific modes of representation and perception. Again, the association between machine and body both made the body a template for the design of machines and made the machine a template for the understanding and cultivation of bodies.

Furthermore, the reciprocity between machine and body means that this process is a circular one. In the previous chapter, I noted that M. Norton Wise has drawn attention to the movement towards automata that depicted 'uncanny or exotic creatures' such as women, animals and racial Others. Wise relates this to a disillusionment with the idea of automata reproducing 'higher' human faculties associated with the white male body, but it also suggests a privileging of the rhythmical and repetitive nature of the automaton's movement. Rhythmic, cyclical, repetitious movements were characteristic of the machines of the Industrial Age, and while machines might thus be opposed to the supposedly more linear and rational productivity of male thought and action, they also evoked a parallel opposition based on gender.

The robot Maria from the 1927 film *Metropolis* is an art deco icon, and embodies the connection between streamlining and the humanisation of technology. Andreas Huyssen, in his influential account of this female robot, sees the desire to reproduce Woman using technology as symbolising a desire to utterly master the nature she represents.

> Clearly the issue here is not just the male's sexual desire for woman. It is the much deeper libidinal desire to create that other, woman, thus depriving it of its otherness. It is the desire to perform this ultimate task which has always eluded technological man. In the drive toward ever greater technical domination of nature, *Metropolis'* master-engineer must attempt to create woman, a being which, according to the male's view, resists technologization by its very 'nature'. (Huyssen 1986: 71)

This is perhaps true, but the association between Woman and Nature is also part of a larger constellation of binaries that includes that dividing mind and body. Woman represents the body as opposed to the mind, and therefore the robot woman is also perhaps an appropriate representative for the nineteenth-century association between machine and body, given its origins in a focus on movement and repetitive labour, and a loss of faith in the mechanical reproduction of 'higher' human faculties.

The depiction of another famous female robot, Hadaly from Auguste Villiers de l'Isle-Adam's *L'Eve future* (1982 (first published 1886)), combines an association between women and the non-cerebral (and, indeed, the decerebration of the love-struck man) with an association between women and artificiality. The fictitious Thomas Edison of the story justifies his creation of a female android on the grounds that the living seductress is already an artificial construction.

> Well then, I thought, if the Artificial, when assimilated to or even amalgamated with human nature, can produce such catastrophes; and since, consequently, any woman of the destructive sort is more or less an Android, either morally or physically – in that case, one artifice for another, why not have the Android herself? (Villiers de l'Isle-Adam 1982: 123)

Hadaly's superiority relative to the woman she mimics is not just the result of better engineering, however; she is innately more appealing precisely because she lacks human biology: '[T]he Android, even in her first beginnings, offers none of the disagreeable impressions that one gets from watching the vital processes of our own organism. In her, everything is rich, ingenious, mysterious' (Villiers de l'Isle-Adam 1982: 130).

Both of these examples therefore suggest that there is a sense in which the android is preferable to the living body due to its freedom from biology, and this idea is most apparent in relation to the idea of a female-gendered robot – hardly surprising given the traditional association of the female body with biology and the 'abject' (Kristeva 1982). Perhaps this also explains David Levy's hope that female robots will one day be available as an alternative to living women as sexual partners.

As such – and contrary to our aesthetic prejudices – these representations can be seen as akin to Michelangelo's *David* and other 'classical' representations of the body, as discussed in the previous chapter. They do not, obviously, subscribe to an idealisation of naturalism, but they do reflect the same values pertaining to the 'civilising' and making 'polite' of the body as discussed there. Where the 'civilising' of the body (in Elias's terms) seeks to hide the reality of its distasteful biological functions and lack of self-containment, Hadaly is considered superior because she has no biology to hide.

What such examples suggest is a nexus of shared concerns and values joining dominant attitudes to both human bodies and machines. Rather than the natural attributes of machines being applied to human bodies, or the natural attributes of human bodies being applied to machines, we can see that both are worked upon in certain ways in an attempt to satisfy certain values, but that those values are, in key areas, shared between the two and reflect a close association between them.

The contrast between fragmentation and unification, then, does not map onto an opposition between body and machine. On the contrary, a tension between fragmentation and unification exists in representations of both, which only further strengthens the sense of interchangeability between them.

Again, the association between machine and body works in both directions: the body is understood to be like a machine, but the machine is also understood to be like a body. During the Renaissance, a view of the body as disassembled mechanism in the anatomy theatre interacted with a view of the body as a unified, self-contained, aestheticised object of admiration. While the body comes to be seen as like machines in being a collection of mechanistic systems, the association of machines with bodies also creates a desire to have machines present themselves as self-contained, aestheticised objects of admiration. Bodies are trained to be polite, neat and beautiful, and machines are increasingly built to reflect the same values. Again, the association of body and machine stimulates this desire, and makes its results more satisfying: looking at the machine as we look at a living body, the perception of qualities we find attractive and endearing in a living body causes us to find the machine also attractive and endearing.

While efforts to create robots, generally arising from science and engineering, have sought to cloak themselves in pragmatic justifications – medical research, replacement of human industrial labour, assistance for the elderly, or even unavoidable preparation for a robot revolution whose arrival has been pre-ordained independently of human wishes or desires – the sheer variety of these justifications, which shift to make themselves relevant to the concerns of particular historical moments, suggests that they are justifications created after the fact rather than direct motivators. Of course, the enthusiasm for such efforts has ebbed and flowed over time, and I must once again make it clear that I am not at all suggesting that human beings have a natural, innate drive to build robots.

Rather, over the past two centuries the further development of mechanist accounts of the body and new, technologised, alienating and objectifying ways of investigating the body have created a deeper, historically and culturally specific connection between human body and machine. This has made the machine not just one more object of human vision to be viewed through an anthropomorphising prism but also something credited with sharing a set of fundamental qualities with human anatomy. The machine's self-movement invites associations with the living body, but scientific discourses that portray the body as self-contained and differentiated from its environment and other bodies make the machine's quality of self-contained organisation seem all the more evocative of the human form.

However, the circulation of these visual cues becomes more and more abstracted and immaterial over time. In a heavily mediated and informaticised society, they can be applied more widely and easily than ever before, and an increasing amount of what we see does not depend upon the kind of physical working upon materiality that makes living bodies or artefacts like cars look the way they do. The move from the Industrial Age to the Information Age has provided new frontiers for such aesthetic principles as mechanistic accounts of the human body have moved from Descartes' levers and pulleys and La Mettrie's cogs and gears to computation. This has taken the scientific abstraction and fragmentation of the body to a new plane entirely: having broken the objectified body down into progressively smaller and smaller parts, the computational approach dematerialises it completely, understanding it as simply arrangements of information. Once again, this view depends entirely upon an analogy between human body and machine, which is once again transformed into that powerful tautology: a metaphor that is literally true (see Channell 1991: 30). When the body is understood in this way, new machines produce new bodies, and it is to the new bodies produced by new information-processing machines that the next chapter will turn.

Chapter 4
Informateriality

The Silver Lady

One of the most ambitious machines of the nineteenth century was Charles Babbage's Difference Engine (Figure 4.1), a device so challenging that Babbage never managed to complete it despite nearly twenty years of work and substantial government funding, although working difference engines were built from Babbage's designs in the twentieth century as historical illustrations.[1] The fact that a movement to realise and thus vindicate Babbage's Difference Engine designs did not arise until over 150 years later illustrates that the value and significance currently attributed to them appeared only relatively recently. The reason for this is obvious: what makes the Difference Engine notable is that, while it was in many ways a product of its age – being a large metal contraption bristling with rotating gears and intended to replace human labour – it was intended to replace mental, rather than physical, exertion. Rather than replacing the muscle power of a galloping horse or a man swinging a hammer or manoeuvring a shuttle through a loom, the Difference Engine replaced the work of the computer, a human worker whose exertions were limited to the mathematical. While most other nineteenth-century machines generally could be understood to direct their power at the taming of the physical environment – hauling loads and people great distances, or stamping, sawing, pounding or weaving new industrial goods – the Difference Engine was intended to crunch numbers. Rather than being a machine that generated energy, the Difference Engine generated what would now be called information, and, like other nineteenth-century technologies that automated and industrialised ephemeral phenomena – for example the cinema – the Difference Engine seems more significant in the Information Age than it did in the Industrial Age that produced it.[2]

Even as he struggled to realise his designs for the original Difference Engine, Babbage was diverted by plans for further machines that would improve upon it – first the Difference Engine No.2, and then successive designs for a more ambitious machine dubbed the Analytical Engine (Babbage 1994: 53–74), which it was hoped could display a 'capacity for *memory* and

---

1   It is also true that devices inspired by Babbage's work were produced before the end of the nineteenth century.

2   However, its material value at the time should not be understated – the mechanised processing of numbers had practical value for the maintenance of the British economy and Empire through its use in areas such as shipping.

Fig. 4.1   A test fragment of Charles Babbage's never-realised Analytical Engine. © Science Museum/Science & Society Picture Library. Reproduced with permission.

anticipation' (Schaffer 1994: 206–7). The Analytical Engine would have been the world's first universal computer and, although it was never built, it did inspire the creation of the world's first computer program, which Ada Lovelace, the daughter of Lord Byron, created in the expectation of it being ultimately fed into the completed Analytical Engine on a punchcard. Babbage's unrealised plans, therefore, have been imbued with a retrospective significance as harbingers of the era of digital computing.

At the same time, of course, Babbage's plans were not – and could not have been – for electronic machines. The Analytical Engine was related to the automata of Jacques de Vaucanson (by way of the automated loom) at least as closely as the digital computer. In fact, in his memoirs, Babbage's fascination with automated machines is embodied in a mechanical woman:

During my boyhood my mother took me to several exhibitions of machinery. I well remember one of them in Hanover Square, by a man who called himself Merlin. I was so greatly interested in it, that the exhibitor remarked the circumstance, and after explaining some of the objects to which the public had access, proposed to my mother to take me up to his workshop, where I should see still more wonderful automata. We accordingly ascended to the attic. There were two uncovered female figures of silver, about twelve inches high.

One of these walked or rather glided along a space of about four feet, when she turned round and went back to her original place. She used an eye-glass occasionally, and bowed frequently, as if recognizing her acquaintances. The motions of her limbs were singularly graceful.

The other silver figure was an admirable *danseuse,* with a bird on the forefinger of her right hand, which wagged its tail, flapped its wings, and opened its beak. This lady attitudinized in a most fascinating manner. Her eyes were full of imagination, and irresistible. These silver figures were the chef-d'oeuvres of the artist: they had cost him years of unwearied labour, and were not even then finished. (Babbage 1994: 12)

After the death of 'Merlin', his collection of mechanical wonders was purchased by a private museum, and the unfinished silver dancer was abandoned in an attic until the museum's contents were in turn auctioned off. Charles Babbage himself then managed to purchase the silver *danseuse* and completed her, putting her on display on a pedestal in his home. The Silver Lady was clearly a seductive figure for Babbage, who referred to her as if she were alive and even had female friends create clothes for her. Babbage noted that visitors to his home admired the Silver Lady more often than they did his display of the unfinished Difference Engine (Babbage 1994: 273–4) – presumably the smooth, curved, performing body of the Silver Lady was more captivating than the disembodied mathematical 'brain' of his own invention.

The Difference Engine gestures forward to a time when the relationship between machine and body would be transformed by the appearance of the computer, despite the fact that the computer displays none of the physical energy and animation that had previously invited an association between machine and body. Just as with clockwork and the motor, the computer inspired a new set of values concerning what was most important about the body, a set of values that the computer, having inspired, also shared. These values were built around the importance of information for both body and machine, and so marginalised the material attributes of both.

Virtualisation

In the 1980s it was already clear to Donna Haraway, author of 'A Cyborg Manifesto' (1991: 149–81), that new digital technologies were having a profound effect on how the human body was understood, arguing as she did that 'we are living through a movement from an organic, industrial society to a polymorphous, information system' (Haraway 1991: 161). Haraway adopted the cyborg as a problematic representative of these changes and the possible embodiment of a positive future, an appropriation that instantly captured the imagination of many readers (see Sofoulis 2002). For Haraway, the cyborg's breaching of boundaries separating human and machine – often understood to be threatening – showed the need to transcend 'the analytic resources developed by progressives [that] have insisted on the necessary domination of technics and recalled us to an imagined organic body to integrate our resistance' (Haraway 1991: 154). Recognising a moment when fundamental categories were being redefined, Haraway urged her readers to wrest some measure of control over the outcome.

Today, however, the cyborg seems a little dated,[3] a dinosaur of the *zeitgeist* that once roamed widely through Hollywood movies and science fiction novels but is now seen relatively rarely. Even by the time of the second Terminator film in 1991, the Schwarzenegger cyborg of the original had been transformed into a kind of faithful puppy, an endearingly clunky piece of obsolete machinery doggedly endeavouring to be of service in the face of threateningly sleek and efficient new technology.

The new relationships between body and technology whose emergence was identified by Haraway have not produced strange new amalgamations as she had hoped, but rather a homogenisation that dissolved everything into a soup of information. The cyborg no longer captures the popular imagination for the same reason that it once did: it represents hybridity, the bringing together of distinct, opposing categories. Today, the idea of a robot arm somehow bolted onto a fleshy torso or a chrome eyeball peering out of a human skull with infrared vision seems clunky and unsophisticated.

When the 1970s television series *The Bionic Woman* was recycled for a short-lived remake in 2007, the eponymous heroine was transformed from a body divided between metal and flesh into a seamless interpenetration of biology and technology: rather than metal servos and wires, she had been colonised by nanomachines, creations that existed on a scale so small as to make distinctions between machines and living cells more-or-less meaningless, and which could move freely inside her fleshy body and interface with it directly.[4] The boundary separating human from machine has already been crossed in popular consciousness, but the result is a monolithic and all-encompassing new term that subjugates categories such as human and machine, rendering them fundamentally interchangeable: information. Even *Terminator 2* re-imagined

---

3 The same could perhaps be said for much of 'A Cyborg Manifesto' itself, given that it was hampered in many ways by the very fact of its prescient engagement with issues that were still in the process of emerging.

4 Nanotechnology will be discussed in detail in the following chapter.

the cyborg threat of the original film as what was, in effect, a purely digital body: whatever internal narrative justification was meant to be secured through references to a 'bio-mimetic alloy', it is clear to the audience that the abilities of the T-1000 terminator were determined by the digital animation effects being used, rather than the digital effects being determined by an independently formulated set of narrative requirements. The second terminator's powers stemmed from the fact that it was a digital entity produced by a computer and sutured into the film.

Here is the breaking down of living organisms into assemblages of information and information-processing devices that Haraway perceptively identified some time ago. But it's hard not to think that, at such an early moment, Haraway could not fully appreciate how final and all-encompassing such a process would be.

The sense that human bodies are becoming saturated or even replaced by information flows has been influentially discussed by N. Katherine Hayles in her work on 'virtuality', which she describes as 'the cultural perception that material objects are interpenetrated by information patterns' (Hayles 1999a: 13–14, Hayles 1998: 69). Hayles points to the wide variety of contemporary experiences that produce such a sense:

> From ATMs to the Internet, from the morphing programs used in *Terminator II* to the sophisticated visualization programs used to guide microsurgeries, information is increasingly perceived as interpenetrating material forms. Especially for users who may not know the material processes involved, the impression is created that pattern is predominant over presence. From here it is a small step to perceiving information as more mobile, more important, more *essential* than material forms. When this impression becomes part of your cultural mindset, you have entered the condition of virtuality. (Hayles 1999a: 19)

Today the Internet, in particular, serves as a seeming confirmation of human virtualisation. The idea that the Internet somehow nullifies our physical bodies and allows us to exist as digital avatars was introduced early to discussions of its effects, and has remained a central theme in celebratory accounts of online interaction. Back in 1995, Sherry Turkle, presumably intoxicated by postmodern theories of identity, made the influential claim that networked interactions were 'affecting our ideas about mind, body, self, and machine' (Turkle 1995: 10): 'In my computer-mediated worlds, the self is multiple, fluid, and constituted in interaction with machine connections; it is made and transformed by language; sexual congress is an exchange of signifiers ...' (Turkle 1995: 15). This claim that the Internet provides a digital space where the Cartesian *cogito* can roam free, living a 'second life' unencumbered by flesh, has lived on, despite the fact that the conceptual problems of this claim were apparent from the start. After all, Turkle's claims were based on the discussion of practices like cybersex or gender performance, and it is hard to understand what either sexual pleasure or desire, or the attribution of gender, could mean without recourse to a physical body.

However, while the dissemination of computers and networks throughout society in the second half of the twentieth century brought with it experiences and discourses that naturalised the idea that the physical body had little – and ever-decreasing – importance, there is still a great deal of continuity between this idea and earlier beliefs concerning the relationship between machines and bodies. For example, while Jonathan Sawday has placed the practice of anatomy in the Renaissance within a broader 'culture of dissection' in which the destructive exploration of the human body was the dominant trope in descriptions of the pursuit of knowledge and understanding (see Chapter 2); Barbara Maria Stafford has pursued many of the same themes through later centuries, and argues that their centrality never went away:

> [W]hat I am terming the anatomical 'method' continues to live on in the twentieth-century informated environment and the nontactile age of the electronic machine. The computer-mediated milieu renders the body nakedly public ... Similarly, one result of the new noninvasive imaging technologies in the area of medicine is the capability of turning a person inside out. If the late nineteenth century developed the photographic sounding of the living interior through endoscopy, gastroscopy, cystoscopy, and, most dramatically, X-rays, the late twentieth century revealed its dark core three-dimensionally through MRI projections. Using radio waves and magnetic fields, this technique for painlessly exploring morphology nonetheless raises the specter of universal diaphaneity. It conjures up foreboding visions of an all-powerful observer who has instant visual access to the anatomy, biochemistry, and physiology of a patient. (Stafford 1991: 48)

The Enlightenment prized the abstract and generalisable as the form of true, transcendent knowledge, producing what Stafford has described as the 'modern theory of abstraction as a sign of superabundant power' (Stafford 1991: 134). Not only the specificities of particular cases, but also the particularities of physicality itself were denigrated or dismissed, resulting in 'the starvation of a fat reality into elegantly thin formulas' (Stafford 1991: 132).

According to this view, true knowledge lay hidden behind the messy, confusing reality presented to our mortal senses, a reality that might even mislead us with false or ambiguous evidence. This idea continues Socrates' belief that knowledge is best attained by the man

> who approaches each object, as far as possible, with the unaided intellect, without taking account of the sense of sight in his thinking, or dragging any other sense into his reckoning – the man who pursues the truth by applying his pure and unadulterated thought to the pure and unadulterated object, cutting

himself off as much as possible from his eyes and ears and virtually all the rest of his body, as an impediment which, if present, prevents the soul from attaining to the truth and clear thinking … (Plato 1993: 126–7)

However, where the philosophers of classical Greece had tried to gain access to truth using little more than the resources of their own minds, new technological devices later promised to free us from the limitations of our unreliable senses, serving as prostheses that could augment the powers of our organs of perception, or even capture information lost to them entirely. The act of dissection itself was a crude technological intervention that opened up new vistas to the human eye and human understanding, allowing us to see what had previously been shrouded in mystery, but in time there would be a procession of new technologies of perception, a procession that, as noted by Stafford above, continues to this day

The practice of anatomy has always tried to straddle – perhaps rather awkwardly – the divide between material reality and specificity on one hand, and generalised, abstracted information on the other. It seeks to mediate between real, physical bodies, and diagrams and general principles, and even negotiate the divide between the animate and inanimate, life and death. In this it is not so different from – and in fact lays the groundwork for – the later investigations of Muybridge and Marey, but of course it does not proceed with such lightness of touch. It doesn't disassemble the flows of light ricocheting from the hide of a galloping horse; rather it sets about disassembling the horse – or the pig, or the monkey or, ideally, the human being. As noted in the preceding chapter, the non-invasive fragmentation of the body achieved by Muybridge, and even the early, invasive investigations of Marey were made possible by the appearance of new technologies. Rather than pulling the body open and turning it inside out so that it was exposed to human sight, new technologies of measurement and image capture allowed investigations more subtle than those relying on unmediated human vision. Believers in the power of Muybridge's images suggested that they might train the eye to see in a way more like the inhuman vision of a machine, but the development of new technologies continued at a pace that outstripped any possible change in the function of human eyes. These machines did the work of transforming human vision in themselves: rather than human sight needing to change, it was only necessary to train that sight on the images produced by machines.

The limits of human visual perception having been reached, investigators began to look at bodies with alien modes of vision that operated outside the visible light spectrum, from the X-ray slide to the magnetic resonance image. This was a chronological progression – new technological discoveries adding new possibilities – but not a process of appearance and obsolescence: today, the X-ray machine and the fMRI machine operate side by side, and bodies are still dissected. Each new view of the human body and its internal processes did not replace those that had come before it; rather it added a new way of seeing, a new representation of ourselves, to the existing collection.

Emblematic of this new vision was the X-ray photograph (see Cartwright 1995: 107ff.). Its particular fascination sprang from the fact that, while producing an utterly novel view of the human body, it worked in much the same way as human vision, and could even be captured with photographic plates like visible light. This made it more understandable in everyday human terms than dissection, or even the altered scales of microscopy or high-speed photography, even while it was astonishingly unfamiliar in its results. As reflected in the 'X-ray vision' of superheroes like the godlike extraterrestrial Superman, the X-ray image in effect recalibrated human sight so that it left the familiar visible light spectrum, transforming it into the vision of some hypothetical alternative species of human being. However, in so doing, it obliterated qualities of human vision such as depth perception, and the dangers of prolonged exposure to X-rays meant that the representation of movement was uncommon.

Over time, the X-ray image was joined by more complex techniques for imaging the body, including those in which computer technology acted to bridge the gap between machine and human vision. As reflected in the name CAT (*computer assisted* tomography) scan, a computer can process a stream of digital data so that a form of machinic perception alien to sight is transformed into something amenable to human vision. The images produced by technologies such as CAT, PET and MRI[5] are simulations of the human body more than reflections of it: the computer generates a visual model based upon data gathered by the scanning machinery, producing something like a personalised anatomical illustration, complete with colour-coding to make it easier to understand. It is, therefore, a kind of automated, non-invasive anatomisation whose development was presumably motivated by the same desires underlying the adoption of technologies like X-ray photography before it. Where once the anatomist pulled the body apart and an artist (or, later, a photographer) interpreted the results to create a visual artefact amenable to human vision, now the process is automated, utilising forms of perception that can see *through* the body so that dissection is unnecessary, and which make inhumanly perfect measurements. This information is then automatically translated into the form of a diagram.

But where is the 'real' body in all this? It is worth noting that, rather than having been lost to us through the increasing mediation and simulation of digital technology, it never seems to have been available in any meaningful way in the first place. Not only was the anatomist's cadaver something not available to the scrutiny of most people, it was, in itself, not a terribly useful representative of the human body as system or machine. The very fact that cadavers must be disassembled – their constituent parts teased apart and isolated – and the results

---

5   See Waldby (2000: 9–11) and van Dijck (2005: 5–6) for a more detailed introduction to many of these new technologies than is necessary here.

of this process then reinterpreted and abstracted through diagrams and other forms of representation in order to become legible, and that even then viewers need special training in order to make sense of the results, highlights the fact that *actual* bodies are simply not terribly amenable to scrutiny from a mechanistic perspective. The physical body has *never been* available to this kind of scrutiny: even at the dawn of anatomy as a discipline, its first set of techniques – those of dissection and illustration – were employed to stage the body, to re-arrange it, to alienate itself from itself and convert it into various artificial representations.

However, the nature of these recreations and refigurings of the body has changed over time. While dissection and early, invasive optical probes (see Waldby 2000: 89–91) created new views of the body by opening up new corporeal planes to the human eye, later technologies provided novel views by presenting the human observer with various kinds of machinic representation: that of the microscope, the camera and the X-ray exposure. More recently still, medical imaging technologies have relied upon computers to construct images from forms of data that are radically incommensurate with human sight, building digital mannikins and models using measurements of invisible properties. They no more constitute 'real' bodies than the deconstructed exiles who stalk the pages of Vesalius's *Fabrica*, but, unlike less sophisticated simulations, they are directly reflective of individuated bodies – in the case of functional MRI, the model even changes state in real time to reflect the moment-to-moment processes of the patient's body.

Such recent representational technologies clearly continue the tradition of bodily representation and refiguring we have seen already at earlier moments, and arise from the marshalling of ever-greater technological resources in an attempt to realise the perfectly objective view of the body at which they had previously hinted. In many ways technologies such as the fMRI simply expand upon or carry further the relationship with the body already seen, but, at the same time, new kinds of technologies inevitably bring new and distinctive characteristics.

In the broadest terms, new digital imaging technologies carry forward the quest to know the human body in some final way, inside and out, achieving an objectivity of perception utterly alien to our usual mode of seeing the human form. In rendering the body transparent to the technician's gaze, in rendering it permeable to a virtual investigator, in rescaling it and even colour-coding it to make it more amenable to a mechanistic and fragmented understanding, it seeks to synthesise a 'perfect' bodily form that combines desirable attributes of dissection, X-ray exposure, textbook diagram, and so on. Ironically and tellingly, the very artificiality of the representation produced in this way underlies its perceived greater truthfulness and accuracy. As a computer-generated simulation of the body, it lacks the messiness, the occlusion, the stubborn materiality that makes real, physical bodies resistant to the search for truth and understanding.

When seeking to understand the functioning of a body as a mechanistic system, such computer-generated bodies are 'better than the real thing'. As a body without biology, the computer-imaged body does not produce fluids that make a mess or hide the object of internal scrutiny; there are no veils or tangles of tissue needing to be cut away or clamped. The individuated body *becomes* an anatomical diagram, its own key and reference, eager to explain itself to the technician who scrutinises it.

Emblematic of this new mode of investigating the body is the Visible Human Project – a virtual anatomist's cadaver created by digitising information about a real corpse – and especially its first subject, the Visible Man (Figure 4.2).[6] Either explicitly or implicitly, the scholarly and popular fascination that greeted this representation of the human body hinged primarily on the idea that the Visible Man was a kind of mascot for a new era in how the body was represented and understood. Certainly, the fact that the subject of this transformation was an executed murderer (Joseph Paul Jernigan, who had donated his body to the project while on death row in Texas (van Dijck 2000: 276)) gave the business an extra fascination, not least because the transubstantiation of a killer into the realm of cyberspace sounded like the plot of a B-grade horror movie.[7] (In fact, the appropriation of an executed murderer's body for anatomical investigation linked the event to the history of public dissection, in which 'anatomisation' was the most fearsome sentence that could be meted out to serious criminals (Egmond 2003: 108–9).)

Being simultaneously annihilated and given a strange kind of immortality by the magical power of digitisation, the parallel between the Visual Man and Moravec's fantasy of technological rebirth is too obvious to require pointing out. However, turning a body into information isn't quite as clean and bloodless at it might as first seem. While Moravec's account of human digitisation might be dispassionate about the destruction of the brain, and not even bother to note the ultimate fate of the dead body discarded at the end of the process, the creators of the Visible Man were required to deal in a practical way with a real-life corpse. After Jernigan's life had been ended by lethal injection,[8] his body was cocooned in a fast-hardening foam and CAT scanned. Next his body was

---

6 For more on the VHP, see, for example, Waldby (2000), Cartwright (1997) and Treichler et al. (1998).

7 Perhaps a cross between Wes Craven's 1989 *Shocker*, a movie about a killer who, after being executed by electric chair, becomes a murderous spectre haunting the power grid, and Russell Crowe's pre-A-list performance as a computer-generated serial killer in the 1995 film *Virtuosity*.

8 Jernigan's original sentence of death by electric chair had been changed to lethal injection so as to leave his corpse in better shape for the process (van Dijck 2000: 276). The fact that this saved Jernigan from the possibility of a painful, lingering death via electrocution was only a fortuitous byproduct of a concern for the state of his remains.

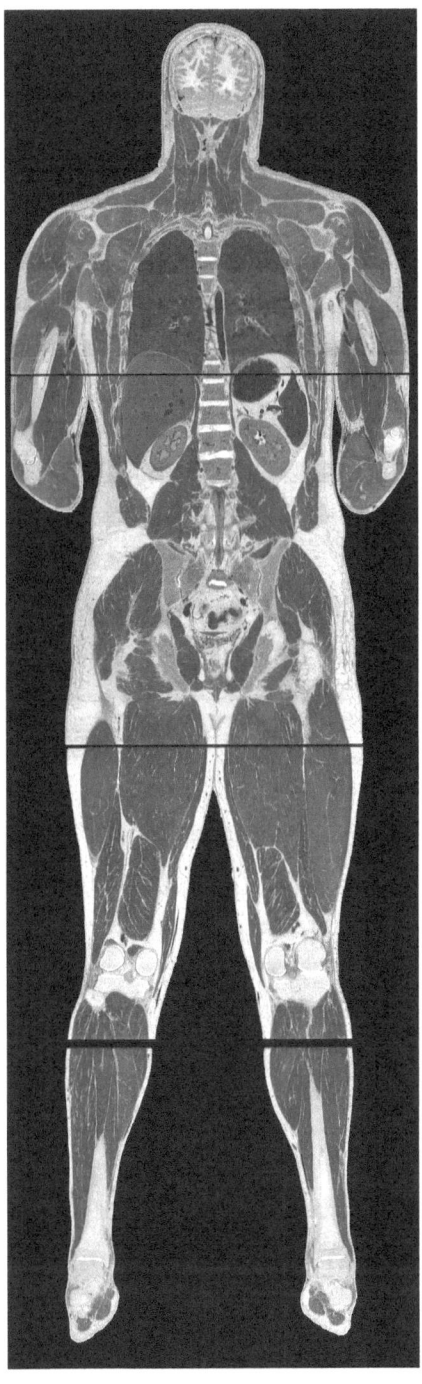

Fig. 4.2   The VHP Visible Man. © Visible Human Project/National Library of Medicine. Reproduced with permission.

frozen and cut into four chunks of a manageable size and embedded in gelatine. These four chunks, frozen to the temperature of dry ice, were then sliced by a specially constructed device called a cryomacrotome into 1,872 slices 1mm thick, each of which was digitally photographed and imaged using a technique that allowed a computer program to recompose these fragments into a unified image of a human body (Spitzer 2002, van Dijck 2000: 275, Waldby 1997: 2). The entire process took nine months to complete.

Broadly speaking, this event, which inaugurated the Visible Human Project's creation of a library of such images (the Visible Man was joined by a Visible Woman in 1995), is simply a continuation of the use of successively more sophisticated technologies to turn the human body into an intelligible system transparently available to human scrutiny. Indeed, as far back as 1907 the same basic approach was used to photograph thin slices of brain and create a film that took viewers on a 'virtual tour' of its interior (Cartwright 1995: 96–7). What differentiates the Visible Man, however, is the fact that his body was destroyed as a physical entity only to be reconstituted as digital data.

It might be argued that the fascination that accompanied the unveiling of the Visible Man depended primarily on the fact that the experience of Joseph Jernigan seemed to capture a sense in the wider population that our own bodies were undergoing a similar process (see Waldby 1997: 1). Today, it seems more fitting to take the Visible Man, rather than the cyborg, as the harbinger and exemplar of technology's problematisation of the boundaries between the natural and the artificial, and this sense is reinforced by the film *Terminator Salvation*, the 2009 attempt to revive the *Terminator* franchise, given that its protagonist is transformed into a cyborg as a result of signing away his body for computer research while awaiting execution in a clear referencing of Joseph Jernigan and the VHP.

It might be argued that it is becoming more and more common for all of us to feel that, both in our everyday experience and in our dominant frameworks for understanding ourselves, we are being dematerialised and transformed into little more than walking accretions of data, and the Visible Man seems to symbolise this process.

New medical techniques for representing the body reflect this, but the role of scientific research into the body extends further. In the nineteenth century, the study of thermodynamics led to the human body being understood as a motor, but in the twentieth century, the study of genetics most importantly produced the belief that it was an information system.

Bodies of Information

The discovery of DNA in the mid-1950s was believed to have unveiled the hidden 'code of life', establishing that the ultimate truth of the living body exists both at an inhuman scale and in the form of information, a kind of blueprint, code or computer program that determines both its physical form and even its behaviour. A number of historical investigations have since highlighted the degree to which our understanding of molecular biology has been shaped by factors arising from its appearance at a particular cultural, technological and scientific moment early in the Cold War (e.g. Kay 2000: 1–11).

Much had been understood about heredity and genetics since the 1930s, but Lily E. Kay, in her authoritative account *Who Wrote the Book of Life?* (2000), argues that 'There were no genetic messages in the 1930s, only biochemical and genetic specificities; genes did not transfer information before the 1950s' (Kay 1997: 29). Although Erwin Schrödinger in the 1940s had proposed that heredity might be transmitted by a mechanism like Morse code, Kay maintains that the idea underlying this concept was fundamentally different from the later understanding of DNA (a molecule then still unknown) as holder of information. This is perhaps illustrated by Schrödinger's conceptualisation of communication in terms of the nineteenth-century technology of the telegraph, rather than the twentieth-century technology of the computer. To quote Kay:

> In the 1950s molecular biology underwent a striking discursive shift: it began to represent itself as a communication science, allied to cybernetics, information theory, and computers. Through the introduction of terms such as *information, feedback, messages, codes, alphabet, words, instructions, texts,* and *programs,* molecular biologists came to view organisms and molecules as information storage and retrieval systems. Heredity came to be conceptualized as contemporary systems of communication, guidance, and control. This linguistic and semiotic repertoire was absent from molecular biology before the 1950s. Based on these scriptural representations, the genome could be read and edited unambiguously by those who know, laying claims to new levels of control over life: beyond control of matter there was now control of information, the word. (Kay 1997: 24–5)

Molecular biology assumed its modern form and influence in the early 1950s, riding a wave of popular fascination and academic excitement generated by Watson and Crick's modelling of the DNA molecule in 1953 (see Chapter 5). Molecular biology, as an emerging field, validated itself by casting its significance in the terms of what was arguably the most fashionable and influential system of thought at the time: information theory. However, in order to do so, it was necessary to take information theory – which had already shifted the concept of information to a substantial degree – and then do further violence to it in order to make it fit the properties of DNA. Kay therefore calls the informatic account of genetics 'a metaphor of a metaphor': 'As a metaphor of a metaphor, the genetic text is a signification without a referent' (Kay 1997: 28). This movement is summarised thus by Evelyn Fox Keller:

[Information] seemed to hold enormous promise for analyzing all sorts of complex systems, even biological systems. Because DNA seemed to function as a linear code, using this notion of information for genetics appeared to be a natural. But as early as 1952, geneticists recognized that the technical definition of *information* simply could not serve for biological information (because it would assign the same amount of information to the DNA of a functioning organism as to a mutant form, however disabling that mutation was). Thus the notion of genetical information that Watson and Crick invoked was not literal but metaphoric. But it was extremely powerful. Although it permitted no quantitative measure, it authorized the expectation – anticipated in the notion of gene action – that biological information does not increase in the course of development: it is already fully contained in the genome. (Keller 1995: 18–19)

The DNA molecule was therefore understood to be the custodian of the informatic stuff of life, and was credited with the highest value and authority relative to the remainder of the human body: 'Everything else – the extranucleic body of the organism – is somatic surplus, designed, constructed, and maintained by the genetic psyche, the all but incidental medium of genetic transmission' (Keller 1995: 97).

This idea reaches its most explicit and extreme expression in Richard Dawkins's famous book *The Selfish Gene* (2006), first published in 1976, in which human beings are cast as 'robot vehicles blindly programmed to preserve the selfish molecules known as genes' (Dawkins 2006: xxi). While this account once again casts the body as an automaton, the soul of Descartes' earlier account has been replaced by colonies of genes that have taken refuge 'safe inside gigantic lumbering robots, sealed off from the outside world, communicating with it by tortuous indirect routes, manipulating it by remote control' (Dawkins 2006: 19–20).[9]

---

9 Such passages have been quoted in an unflattering light often enough that Dawkins felt obliged to include a defence of them in the thirtieth anniversary edition of the book, where he says:

> Part of the problem lies with the popular, but erroneous, associations of the word 'robot'. We are in the golden age of electronics, and robots are no longer rigidly inflexible morons but are capable of learning, intelligence, and creativity. Ironically, even as long ago as 1920 when Karel Capek coined the word, 'robots' were mechanical beings that ended up with human feelings, like falling in love. People who think that robots are by definition more 'deterministic' than human beings are muddled (unless they are religious, in which case they might consistently hold that humans have some divine gift of free will denied to mere machines). If, like most of the critics of my 'lumbering robot' passage, you are not religious, then face up to the following question. What on earth do you think you are, if not a robot, albeit a very complicated one? (Dawkins 2006: 270–71)

Key to the larger confluence of ideas to which the meeting of molecular biology and information theory belongs was the influential but short-lived reign of cybernetics. When figures such as Norbert Wiener, Claude Shannon, Warren Weaver and John von Neumann developed and elaborated the principles of cybernetics and information theory in the 1950s, these ideas achieved a degree and breadth of influence that has certainly never been seen since, and had perhaps never been seen before. In the intellectual milieu of academia, where disciplines tend to be siloed and exchange between the sciences and the humanities is rare, cybernetics and information theory came to permeate an amazing variety of research pursuits.

Cybernetics is widely known to have been influenced by projects relating to the development of particular kinds of military hardware (most notably Wiener's failed 'AA predictor' (Galison 1994: 229)), and information theory most directly originates in the context of new kinds of technology related to what today we would call 'information processing'. But there followed a kind of mania for these ideas that resulted in attempts to insert them into all kinds of research; just how successful or strained, and how mutated or mangled these principles became in their new homes, varied considerably depending on where they wound up.

> [N]early every discipline in the social sciences (sociology, psychology, anthropology, political science, and economics) as well as the life sciences (immunology, endocrinology, embryology, physiology, neuroscience, evolutionary biology, ecology, and molecular genetics) flirted with the seductive ideas of cybernetics and information theory, with different degrees of productivity and commitment. (Kay 1997: 26)

In the early 1970s, prominent cyberneticist Stafford Beer was even commissioned by the Allende government to create a cybernetic control system that could automatically steer the entire Chilean economy (Pickering 2004: 37), although the CIA-backed overthrow of Allende put an end to work on the project before its efficacy could be put to the test.

Aside from the fact that Čapek's robots were not mechanical beings, this defence is a disingenuous one; if Dawkins did not understand the word 'robot' in this pejorative sense, he surely would not have used words such as 'lumbering' and 'programmed' to describe them; furthermore, if he had truly intended to evoke capacities such as 'learning', 'intelligence' and 'creativity', it's unclear why he would have employed the robot metaphor at all – as these are most obviously and notably human attributes, it would have been wilfully obfuscatory to describe human beings as robots in that context. By the time he arrives at his question, 'What on earth do you think you are, if not a robot, albeit a very complicated one?', it is no longer clear what definition of 'robot' Dawkins could possibly be claiming to use.

The concerns of cybernetics are well summarised by the title of Norbert Wiener's famous book *Cybernetics: Or Control and Communication in the Animal and Machine* (1961). In laying out the discipline's approach, Wiener explicitly positioned it at the end of a history of productive interactions between machines and living bodies, something appropriate for a new technological era. The period beginning with the end of World War II was 'as truly the age of servomechanisms as the nineteenth century was the age of the steam engine or the eighteenth century the age of the clock' (Wiener 1961: 43).

> [T]he many automata of the present age are coupled to the outside world both for the reception of impressions and for the performance of actions. They contain sense organs, effectors, and the equivalent of a nervous system to integrate the transfer of information from the one to the other. They lend themselves very well to description in physiological terms. It is scarcely a miracle that they can be subsumed under one theory with the mechanisms of physiology. (Wiener 1961: 38–43)

In its belief that living bodies and machines both could be understood most importantly as self-regulating systems that used feedback loops of information to maintain equilibrium, cybernetics provided a common language for discussing bodies and machines analogous to, but with greater scope than, the nineteenth-century language of energy. In the words of Peter Galison, 'Where Darwin had assiduously tracked the similarities between human and animal in order to blur the boundary between them, Wiener's efforts were devoted to effacing the distinction between human and machine. Darwin's dog suffered remorse; Wiener's AA predictor had foresight' (Galison 1994: 245–6).

Cybernetics' union of machine and body is well represented by its enduring bequest to popular culture, the cyborg itself – the 'cybernetic organism' originally understood to 'deliberately incorporate ... exogenous components extending the self-regulatory control function of the organism in order to adapt it to new environments' (Clynes & Kline 1960: 27) in order to survive the strains of space exploration. But the biological and mechanical syntheses of cybernetics reached further. DNA became an information-processing system sending out commands that controlled the development of the organism, and the information switching mechanisms of the day provided a model for the human nervous system (Wiener 1961: 116–32). In the words of John Johnston,

> Cybernetic discourse ... tended to speak of machines in terms of living organisms and living organisms in terms of machines. There was in fact an assumed or implicit agreement that the two differed only in the complexity of their respective organization. The unspoken – and perhaps unspeakable – objective directly follows: to bridge the gap between the organic and the inorganic, the natural and the artificial. The cyberneticists' use of the word *automaton* – or,

more often, the plural *automata* – also points in this direction. Conventionally, of course, the term designates a self-moving machine, often a mechanical figure or contrivance meant to convey the illusion of autonomy. The cyberneticists, however, speak of natural *or* artificial automata, of automata 'in the metal or in the flesh,' as Wiener puts it … (Johnston 2008: 29)

Following from the work of authors such as Kay and Keller, N. Katherine Hayles's interest in cybernetics results from a sense that it initiated an ongoing trend of disregarding the importance of human embodiment, and understanding a wide variety of phenomena as composed of immaterial information flows. However, Johnston has more recently highlighted the important differences between the cybernetic view and the work on information and computation that followed it (see Johnston 2008: 60–61). Because cybernetics was centrally concerned with feedback and self-regulation, interactions between the organism and its environment were crucial to its understanding of the cybernetic system; consequently, and contrary to the charges of dematerialisation and disembodiment brought by Hayles, cybernetics was very much concerned with material interaction and the behaviour of bodies, even if those bodies were understood to function according to informatic principles (Johnston 2008: 30). As a result, cybernetic research often took the form of attempts to model or reproduce bodies. Because bodies were understood to be most importantly self-regulating systems, these artificial bodies were far removed from the automata of the eighteenth century, but they were similar in that they represented an attempt to understand what were considered key attributes of the living body by recreating them in a mechanical system. Cybernetic automata such as W. Grey Walter's mechanical 'tortoises', Elmer and Elsie (Johnston 2008: 49–53), were embodied agents that interacted with their environment and each other to produce (relatively) complex behaviour.

However, the reign of cybernetics was relatively short, and it might be argued that it fell from fashion precisely because did not make the leap to a disembodied and dematerialised view of information; Johnston has suggested that cyberneticists' more embodied, materialist view of information flows was an inevitable result of the machines they used as their models (Johnston 2008: 59). The early computers with which the first generation of cyberneticists was familiar were undeniably physical artefacts: they were huge and needed the constant physical ministrations of technicians, and programs were physically instantiated in the world as punch cards. The influence of cybernetics waned with the rise of a contemporary understanding of the computer as simply a generic, replaceable piece of hardware running immaterial and privileged software. Cybernetics was elbowed aside by a new set of ideas that arose from this new view of computation, and which pushed the idea of the machine as a model of the nervous system into the background, focusing instead on the idea that human consciousness was like a computer program. The body–machine connection therefore evolved towards a belief that *literal* similarities between brain and computer – while still widely believed to exist – were less important than the idea that both systems processed information in an equivalent way. The mechanical bodies built by cyberneticists, like Elmer and Elsie the mechanical tortoises, were supplanted by the (at least hypothetical) disembodied symbol processors of AI, which 'essentially reduced artificial intelligence to software and reinstalled a Cartesian duality that cybernetics – at least at its best moments – had entirely transcended' (Johnston 2008: 61).

Computationalism

With the fall from favour of cybernetics, the development of a focus on abstract information continued to develop, most importantly in two areas. The first, as already noted, was the idea of DNA as an information system. The other was the ongoing belief in an interchangeability between computers and the human brain. This second area produced the broad grouping of cognitive science, which brings together neuroscience, robotics, Artificial Intelligence (AI), philosophy, clinical psychology and more under a common banner. Within this body of work, it is not just possible, but is in fact the norm, to understand brains with reference to computers, and computers with reference to brains.

As indicated above, central to this convergence of thought is a belief that the human brain is a computational system. Consciousness, or the mind, is a piece of software running on the hardware of the brain (Schwindt 2008: 7, Foerst 1999: 379). As expressed by the philosopher David Chalmers,

> Perhaps no concept is more central to the foundations of modern cognitive science than that of computation. The ambitions of artificial intelligence rest on a computational framework, and in other areas of cognitive science, models of cognitive processes are most frequently cast in computational terms. The foundational role of computation can be expressed in two basic theses. First, underlying the belief in the possibility of artificial intelligence there is a thesis of *computational sufficiency*, stating that the right kind of computational structure suffices for the possession of a mind, and for the possession of a wide variety of mental properties. Second, facilitating the progress of cognitive science more generally there is a thesis of *computational explanation*, stating that computation provides a general framework for the explanation of cognitive processes and of behavior. (Chalmers 2011: 324)

This idea that brains and computers work in a fundamentally similar way appears very rapidly after the appearance of universal computing, which might be surprising to modern observers, for whom early computers might seem far too clunky and limited to invite comparisons with the human brain. With the benefit of hindsight, it might be argued that the assertion of this commonality owed a fair amount to an early lack of clarity concerning just what the development of computers could eventually achieve, and a lot more to the fact that little was known about how the human brain itself functioned. To provide a sense of this early confidence, I will quote from Daniel Crevier's 1993 history of the AI movement at length:

> After Herbert Simon's celebrated forecast in 1957 that 'in a visible future [machines will handle problems in a range] coextensive to the range to which the human mind has been applied,' he went on to predict:
>
> 1. That within ten years, a digital computer will be the world's chess champion, unless the rules bar it from competition.
> 2. That within ten years a digital computer will discover and prove an important new mathematical theorem.
> [...]
> 4. That within ten years most theories in psychology will take the form of computer programs, or of qualitative statements about the characteristics of computer programs.
>
> With the public reading of these lines from a paper by himself and Allen Newell, the future Nobel laureate set the boastful tone that marked the relationship between AI and the news media for many years afterward. One year later, Newell, Shaw, and Simon reiterated these estimates even more optimistically: 'In another place, we have predicted that within ten years a computer will discover and prove an important mathematical theorem. On the basis of our [recent] experience with the heuristics of logic and chess, we are willing to add the further prediction that only moderate extrapolation is required from the capacities of programs already in existence to achieve the additional problem-solving power needed for such simulation.' Only slightly abashed seven years later, in 1965, Simon still maintained that 'machines will be capable, within twenty years, of doing any work that a man can do.' And in a 1967 book, Marvin Minsky proposed a substantially similar forecast: 'Within a generation ... few compartments of intellect will remain outside the machine's realm – the problem of creating "artificial intelligence" will be substantially solved.' John McCarthy, for his part, founded the Stanford AI Project in 1963 'with the then-plausible goal of building a fully intelligent machine in a decade.' (Crevier 1993: 108–9)[10]

For Artificial Intelligence researchers of the 1950s and 1960s, the equivalence between computer and brain was believed to be so direct and comprehensive that building a thinking machine seemed a relatively straightforward task, one that a team of clever people could knock over in a few years. The brain is understood to be a machine that performs a certain function, and – accepting

---

10 All ellipses aside from that at point 3 are Crevier's.

that a sufficiently ingenious engineer should be able to reproduce the function of any existing mechanism – human beings are therefore able to create a mechanism that performs the same function, and is therefore also a brain. As expressed by David Chalmers,

> [S]ystems that duplicate our functional organization will be conscious even if they are made of silicon, constructed out of water-pipes, or instantiated in an entire population … [B]iochemical and other non-organizational properties are at best indirectly relevant to the instantiation of experience, relevant only insofar as they play a role in determining functional organization. (Chalmers 2003: 253)

In this view, 'brain' becomes an abstract mode of description rather than a materially specific object; there is a *quality* of being a brain that human brains have but which might also be expressed by other things entirely. It becomes a kind of Platonic form, devaluing the matter of actual biological brains. The brain is most importantly understood as a function, not a physical organ.

This belief arose in the same way as La Mettrie's belief that another Jacques de Vaucanson could create a thinking, speaking clockwork automaton. If we answer the question, 'How does the human body work?' with the answer, 'Like a clockwork machine,' then our conception of how the human body works is likely to have few if any features we could not imagine being instantiated in clockwork. Similarly, if we answer the question, 'How does the mind work?' with the answer, 'Like a computer,' then our conception of the mind is likely to have few if any features that cannot be reduced to computation. Once again we have a literalising of metaphor, a reciprocity in which each term simultaneously serves as a metaphor for the other with neither functioning as a point of origin. The brain is like a computer is like a brain is like a computer … On and on in an infinite spiral. The result is a situation in which robots can become a model for understanding human beings, rather than *vice versa*.

> Our body is analogous to the robot's mechanical body. Our brain is analogous to the robot's computer hardware including the physical hard drives which contain its control software. Our mind is analogous to the computational processes that result by the running of this control software on the robot. Our first-person conscious experiences are analogous to the particular stream of tokened data structures that the robot uses to model its own perceptions, goals, and actions as part of its self-model (used as a higher-order guide for reasoning about and planning its future behaviors) … Although in principle there may be some deep flaw in this analogy between a human and a mechanical robot, it is an analogy that rests at the core of all of our research in cognitive science, neuroscience, and even biology itself. (Hayworth 2012: 89–90)

Back in 1950, in the article 'Computing Machinery and Intelligence,' the father of the universal computer, Alan Turing, already had posed the question,

'Can machines think?', and answered, 'I believe that at the end of the century the use of words and general educated opinion will have altered so much that one will be able to speak of machines thinking without expecting to be contradicted' (Turing 1950: 442). In the same article, Turing gives serious thought to the possibility of cultivating thinking computers with human-like abilities, for example by teaching them like children. But his most enduring proposal was for what he called the 'imitation game', but which has since become famous as 'the Turing Test':

> It is played with three people, a man (A), a woman (B), and an interrogator (C) who may be of either sex. The interrogator stays in a room apart from the other two. The object of the game for the interrogator is to determine which of the other two is the man and which is the woman. He knows them by labels X and Y, and at the end of the game he says either 'X is A and Y is B' or 'X is B and Y is A'. The interrogator is allowed to put questions to A and B thus:
>
> C: Will X please tell me the length of his or her hair? Now suppose X is actually A, then A must answer. It is A's object in the game to try and cause C to make the wrong identification. His answer might therefore be
>
> 'My hair is shingled, and the longest strands are about nine inches long.'
>
> In order that tones of voice may not help the interrogator the answers should be written, or better still, typewritten. The ideal arrangement is to have a teleprinter communicating between the two rooms. Alternatively the question and answers can be repeated by an intermediary. The object of the game for the third player (B) is to help the interrogator. The best strategy for her is probably to give truthful answers. She can add such things as 'I am the woman, don't listen to him!' to her answers, but it will avail nothing as the man can make similar remarks.
>
> We now ask the question, 'What will happen when a machine takes the part of A in this game?' Will the interrogator decide wrongly as often when the game is played like this as he does when the game is played between a man and a woman? These questions replace our original, 'Can machines think?' (Turing 1950: 433–4)

Paradoxically, in this test the machine can only prove that it is the same as a human being if the judge believes that human beings and machines cannot be the same; if the judge believes this, the machine can then go on to demonstrate qualities that the judge believes mark an irreconcilable difference between human and machine. If the judge genuinely believes in an equivalence between human and machine intelligence, on the other hand, then presumably the judge will not be able to formulate any questions believed capable of catching the machine out.

The Turing Test has been extremely influential. Since 1990 there has even been the Loebner Prize, a competition that seeks to put the Turing Test into effect (although the philosopher Daniel Dennett resigned his position as chairman of its prize committee in disgust after three competitions, clearly feeling that it was not sufficiently intellectually rigorous (Dennett 1998: 27–9)). The perceived value of the Turing Test is clear: as Turing intended, it sidesteps the question of what the intelligence referred to by 'artificial intelligence' actually is – a difficult enough question to answer concerning human beings, let alone machines. The brain is 'black boxed': certain kinds of information are understood to go in and certain kinds of information are understood to come out, but what happens in between can remain a mystery. If you can build some other kind of box that produces the same kinds of outputs in response to the same kinds of inputs as a brain, then you have the right to call it a brain. This behaviourist approach to intelligence has been necessary for AI, given its problematic goal of creating a particular phenomenon (intelligence) despite not knowing what that phenomenon actually is (see Riskin 2007: 10). Treating the brain as a black box has allowed AI to proceed without an understanding of what things like human intelligence or consciousness are, or how living brains function.

The other advantage of the Turing Test, as Turing himself remarked, is that it allows researchers to dispense with 'irrelevant' factors such as bodies and focus simply on the mind:

> The new problem has the advantage of drawing a fairly sharp line between the physical and the intellectual capacities of a man. No engineer or chemist claims to be able to produce a material which is indistinguishable from the human skin. It is possible that at some time this might be done, but even supposing this invention available we should feel there was little point in trying to make a 'thinking machine' more human by dressing it up in such artificial flesh. (Turing 1950: 434)

For Turing, human intelligence is unambiguously understood to be immaterial and disembodied, a Cartesian *cogito*, for whom the presence or absence of flesh is merely incidental. However, there are two ironies in this statement. First, the Turing Test itself, rather than being a purely intellectual challenge – like a mathematical problem for example – is actually a test of what might be termed 'social intelligence'. To pass the test, a machine would need to understand and play upon the judge's understanding of categories of identity and kinds of interpersonal interaction that are dependent upon bodies. In the initial description, the challenge is to differentiate two subjects according to gender, making reference to factors such as physical appearance, and the machine would play in an analogous way by seeking to gain the judge's trust and convince the judge of their shared humanity. The second irony is that Descartes' formulation has been reconfigured so that, instead of a machine body differentiated from the immaterial realm of thought, Turing refers to a machine mind independent of a body that is, if not literally dematerialised, then certainly conceptually dematerialised by its lack of importance or influence.

In her book *Affect & Artificial Intelligence* (2010), Elizabeth A. Wilson gives a great deal of attention to 'Computing Machinery and Intelligence', arguing that it betrays a duality in Alan Turing's speculation regarding thinking machines. At one point he expresses a belief that machine intelligence might be manifested in relation to 'a very abstract activity, like the playing of chess' (Turing 1950: 460); later, when considering how a machine might be imbued with the capacity for thought, he considers the possibility of making a child-like machine, which could then be taught. Wilson uses these statements to argue that Turing was interested both in abstract models of thought and those more mindful of embodied engagement with the environment, represented by 'chess (abstraction) and the child (sensate embodiment)' (Wilson 2010: 22) respectively. However, Turing's understanding of childhood learning is clearly very different from Wilson's, and he quite specifically excludes sensate embodiment from it. Regarding embodiment, Turing has already stated by this point that there would be 'little point in trying to make a "thinking machine" more human by dressing it up in … artificial flesh'; regarding sensation, Turing believes that senses play a role only as avenues for the input of information, and one interface is as good as another:

> We need not be too concerned about the legs, eyes, etc. The example of Miss Helen Keller[11] shows that education can take place provided that communication in both directions between teacher and pupil can take place by some means or other. (Turing 1950: 456)

While it is true that Turing had earlier speculated about creating a more fulsome reproduction of the human body and sensorium (Wilson 2002: 38–9), here he specifies that such thoroughness is unnecessary: a direct exchange of information between teacher and pupil is all Turing's model needs. Wilson takes Turing's model of childhood learning to be about embodied, self-directed exploration, but Turing is quite clearly speaking about something very different, perhaps even oppositional to it. According to Turing, 'the child-brain is something like a note-book as one buys it from the stationers' (Turing 1950: 456), a blank slate to be filled with information by an authority figure through '[t]he use of punishments and rewards', although the child-machine will also require 'some other "unemotional" channels of communication' for its education to be successful (Turing 1950: 457). In other words, for Turing, the education of the child functions as an analogy for the transfer of abstract information directly into the 'brain', where it can then be processed by a disembodied intellect.

Using the computer as a model, it came to be widely believed that the brain was a machine that dealt in symbols or tokens representing objects or concepts.

---

11   Helen Keller famously was educated in the late nineteenth century despite being deaf and blind.

For example, if I see a cat, this input is translated into a symbol in my mind, which can then perform operations on it, such as devising a plan to pat the cat, based on the rules and concepts I already have stored in my brain that are associated with the cat symbol. Having done this, the instructions necessary to pat the cat can be outputted to my body to produce that action. The brain is constantly receiving inputs that are translated into symbols, performing operations on those symbols, and then producing outputs.

All that was necessary was a machine that could gather the right inputs, perform the right symbol processing, and produce the right outputs. One of the most celebrated results of this approach was the Stanford Research Institute's ARPA-funded robot 'Shakey' (Nilsson 1984). Developed in the 1960s, Shakey acted on complex commands by building up detailed representations of its environment, which were then used by an executive layer to formulate programs of action that were then passed down to the machinery to be put into action. However, this way of formulating behaviour turned out to be extremely slow and laborious; even in a greatly simplified environment with constrained lighting, colours and shapes, Shakey required hours to produce behaviours that are near-instantaneous in a human being, or even an insect (Grant 1990).

While figures such as Hans Moravec nevertheless continue to hold a traditional view of AI and robotics, maintaining – like other researchers before them – that their inability to realise the early promises made by their predecessors simply results from insufficient computer power (a problem that will inevitably be overcome by the march of technology (Moravec 1999: 52–7)), frustration with the results of creations like Shakey has also produced attempts to formulate alternative approaches. Most famous is the behaviour-based robotics of Moravec's one-time student of Rodney Brooks,[12] who went on to head the Artificial Intelligence Lab at MIT (Brooks 1999: vii). Brooks explicitly rejected traditional AI's focus on disembodied intelligence and internal representations, seeking to build progressively more complex autonomous robots that interact with their environments. From relatively simple (but flexible and robust) insect-like 'creatures' to robots such as the more human-like Cog (see Dennett 1998) and the endearingly sociable Kismet (Breazeal 2002), Brooks and his colleagues have sought to create robots that are both situated and embodied:

> A *situated* creature or robot is one that is embedded in the world, and which does not deal with abstract descriptions, but through its sensors with the here and now of the world, which directly influences the behavior of the creature.
>
> An *embodied* creature or robot is one that has a physical body and experiences the world, at least in part, directly through the influence of the world on that body. A more specialized type of embodiment occurs when the full extent of the creature is contained within that body. (Brooks 2002: 51–2)

---

12   In addition to casting doubts on Moravec's approach to AI, Brooks is also on record as dismissing Moravec's belief in computerised immortality (Brooks 2002: 204–6).

The ongoing disappointments of traditional approaches to artificial intelligence have produced other alternatives, too, such as the Artificial Life (or ALife) movement (Langton 1996) and research into affective computing (Suchman 2007: 232). All arise from a belief that the field of AI has been held back by uninterrogated assumptions concerning what constitutes bodies and minds.[13]

Their focus on embodiment, while an explicit rejection of the idea that human subjectivity is entirely a matter of symbol processing in the brain, obviously is not a rejection of the larger idea that machines and bodies have a fundamental commonality. They simply reject the excesses of the mind-as-information approach by returning to a more direct focus on the body. The ALife movement claims continuity with Enlightenment automata like Vaucanson's famous duck (which is sometimes used as its emblem), and behaviour-based robots are historically most closely related to the feedback-based mechanical creatures of cybernetics; as a result, such approaches to some degree hark back to a pre-computationalist era when the body as physical machine was the primary concern. The philosophical underpinnings of these more recent movements might be quite different from those of the past, not to mention the terms in which they understand the living body to be a mechanistic system, but the link between body and machine remains.

In fact, recent approaches to robotics and AI are increasingly being incorporated into research agendas in which machines are considered to be interchangeable with living bodies. Evelyn Fox Keller has already drawn attention to the ways in which robots such as Cog and Kismet have been designed using models of human infant development, and robots are also being utilised in therapies and evaluations of human functioning (particularly with the successful use of robots as interactive partners for autistic children) (Keller 2007). Of course, cognitive science more generally continues to be characterised by a belief in a set of productive reciprocal relationships between machines and living bodies.

Timothy Lenoir has cited such developments to argue that criticism of AI and robotics research on the grounds that it devalues the body (of the kind associated with Haraway and Hayles, for example) is no longer necessary (Lenoir 2007). However, while such developments certainly highlight the limitations of a disembodied, abstract understanding of the mind, there is still a great deal of variation in what precisely those limitations are understood to be and how it is proposed they be addressed (see Ziemke 2007). As noted by Mark Johnson,

---

13   The significance of these alternative approaches, and others like them, deserves a more extended discussion, but unfortunately constraints of space prevent me from doing them justice. A thoughtful and comprehensive discussion can be found in (Johnston 2008).

Embodiment theory is now well supported by research in the cognitive sciences, yet there remains considerable debate as to what exactly the term *embodiment* might mean … Is the 'body' merely a physical, causally determined entity? Is it a set of organic processes? Is it a felt experience of sensation and movement? Or is it a socially constructed artifact? (Johnson 2007: 119)

In any event, there remains a great deal of thinking and research that continues uninformed by such reappraisals, or works to reconcile them with more traditional informatic accounts of mind and body. Perhaps there is no better illustration of the continued health of the old disembodied, computationalist understanding than the way in which Moravec's mind uploading idea has thrived since he first put it forward, spreading to produce, not only serious scientific research and speculation, but an almost religious wider following.

Ray Kurzweil's New Body

Ray Kurzweil is a famous entrepreneur, technology developer and futurist. His name is most associated with the Kurzweil synthesiser, but he has created numerous other computer programs and devices, such as the Kurzweil Reading Machine, which was the first device to allow the blind to hear print converted to speech, and in 2012 he was given a full-time 'engineering director' position at Google.

Kurzweil has also written several books, which have shifted focus over time from celebrating recent technological developments to speculating – in what some might describe as an increasingly wild fashion – about possible future developments, and today he's known largely as a futurist, his predictions for the future becoming more and more focused on the possibility of human immortality as he has gotten older.

In 2004, Kurzweil even co-published a book entitled *Fantastic Voyage: Live Long Enough to Live Forever*, in which he claimed that, if you could stick around for another fifty years or so, you'd live to see technologies that will make biology, and therefore mortality, a thing of the past. This would seem to be a bit of a good news/bad news story for Kurzweil himself, given that he is now in his sixties; he's clearly at an age where the defeat of death has a special fascination for him, but also one that makes another fifty years of life seem unlikely. However, Kurzweil is clearly determined to make it, and the book is primarily concerned with strategies and therapies he believes can extend one's life long enough to see in the era of immortality. Kurzweil supposedly ingests over 250 nutritional supplement pills a day, and receives half-a-dozen intravenous supplements a week, in an effort to extend his lifespan (Kurzweil 2005: 260).

Evaluating the efficacy of these treatments is beyond my expertise, but I'm not particularly interested in Kurzweil's strategy for surviving until a time when biology becomes obsolete; rather, I'm interested in his certainty that this time will come at all. Kurzweil believes that immortality will come in the form of what he has referred to as 'brain porting' (Kurzweil 1999: 128) – in other words, the mind uploading scenario previously put forward by Hans Moravec – and his faith that the technology required to effect this process will soon arrive comes from his belief in the approach of a revolutionary transformation of human history called the 'Technological Singularity'. Singularitarianism is a brand of technophiliac millenarianism of which Kurzweil is a key apostle (Grossman 2011), having argued for its inevitability in books (Kurzweil 2005), a film documentary (Ptolemy 2009), a feature in *Time* magazine (Grossman 2011), and even through a Singularity University (Anon. 2011).

The idea of a technological singularity is widely believed to have its earliest beginnings in a 1966 suggestion by the famous British statistician and mathematician Irving John Good that, 'It is more probable than not that, within the twentieth century, an ultraintelligent machine will be built and that it will be the last invention that man need make, since it will lead to an "intelligence explosion". This will transform society in an unimaginable way' (Good 1966: 78). However, it was first explicitly put forward by academic and science fiction author Vernor Vinge in the 1980s, and was popularised by a 1993 article in which he suggested that artificial intelligence and the technological amplification of human intelligence would one day unleash exponential progress (Vinge 1993). For true believers such as Ray Kurzweil, however, the Singularity is understood more broadly:

What is the Singularity? From my perspective, the Singularity is a future period during which the pace of technological change will be so fast and far-reaching that human existence on this planet will be irreversibly altered. We will combine our brain power – the knowledge, skills, and personality quirks that make us human – with our computer power in order to think, reason, communicate, and create in ways we can scarcely even contemplate today.

This merger of man and machine, coupled with the sudden explosion in machine intelligence and rapid innovation in gene research and nanotechnology, will result in a world where there is no distinction between the biological and the mechanical, or between physical and virtual reality. These technological revolutions will allow us to transcend our frail bodies with all their limitations. Illness, as we know it, will be eradicated. Through the use of nanotechnology, we will be able to manufacture almost any physical product upon demand, world hunger and poverty will be solved, and pollution will vanish. Human existence will undergo a quantum leap in evolution. We will be able to live as long as we choose. The coming into being of such a world is, in essence, the Singularity. (Kurzweil 2006: 39)

The Singularity is therefore understood to be a moment when the relationship between bodies and machines will be fundamentally transformed, reaching a final point of communion. It therefore represents a final apotheosis for the mechanist understanding of living bodies, one where distinctions between living bodies and fabricated machines disappear entirely. This final disappearance is believed to be possible – and even inescapable – because it rests on the idea that human bodies can be entirely reduced to patterns of information.

For Kurzweil, this transformation will be part of the realisation of the 'intelligent destiny of the cosmos' (Kurzweil 2005: 342–67), and he looks forward to a future when all matter is merged into a computational system with godlike potential. Hans Moravec has made similar science-fictional predictions about a coming age of cosmic information processing (Moravec 1999: 164–8), and this sense of destiny is created by imbuing the principle known as Moore's Law with a cosmic significance. This 'law' began as little more than an observation by Intel co-founder Gordon E. Moore in 1965 when, in commenting on the future of integrated circuits, he noted that the number of transistors that could be fitted on a circuit had approximately doubled each year to that point, and similar growth could be expected to continue for some time into the future (see Kurzweil 1999: 20–21). However, this hardly constitutes a 'law' in any meaningful sense; it's clear that Moore himself was not claiming this to be a law or guarantee of future developments, and no reasonable observer could really feel otherwise. Nonetheless, since the 1960s Moore's Law has been widely mythologised, its details altered to fit historical data, and it is now often claimed to predict increases in computing power – something it originally made no mention of. For his part, Ray Kurzweil has elevated it to the status of a fundamental principle of reality.[14]

Kurzweil not only accepts Moore's Law as an underlying physical property of the Universe, but argues that this exponential rise in computing power began, not in the 1960s with Moore, but with the first computing machines in the early 1900s (Kurzweil 1999: 18–25). He then goes so far as to see this line of technological development as part of a larger cosmic principle, which he calls 'the law of time and chaos': 'In a process, the time interval between salient events (that is, events that change the nature of the process, or significantly affect the future of the process) expands or contracts along with the amount of chaos' (Kurzweil 1999: 29).

According to Kurzweil, this principle explains why changes to the physical properties of the Universe have 'slowed down' over time, while evolutionary and technical changes have 'sped up' over time. Consequently, according to Kurzweil, Moore's Law is in fact driven by a fundamental cosmic law, making the appearance of thinking computers not only inevitable, but part of some

---

14   Ikka Tuomi has presented extended analyses intended to debunk both faith in Moore's Law, and Kurzweil's belief in a 'law of accelerating returns' (Tuomi 2002, 2003).

larger cosmic destiny. As a result, and appropriately enough for someone who has written a book entitled *The Age of Spiritual Machines* (1999), which posits an inevitable destiny for humankind and discusses the transmigration of the soul and immortality, Kurzweil's writing is surprisingly like a religious treatise despite its tech-savvy language.

According to Kurzweil's account, information is an underlying, universal property of all reality, organising the history of the entire Universe according to its laws. There is no fundamental problem with Kurzweil's finding a pattern or overall trend in history along these lines – the world around us is filled with patterns. What is a problem, however, is that Kurzweil concludes that this pattern establishes the existence of a cosmic law, suggesting that there is a pre-ordained metanarrative and inescapable teleology to history. It is reasonable to remark that evolution has produced ever more complex organisms; a new species arising through the mutation of a complex species is itself likely to be complex – it's no accident that chimpanzees split to produce the bonobo, not a new form of bacterium. Similarly, human technology generally (but not always) produces ever more complex and sophisticated devices, building upon the ever-expanding store of technical knowledge and experience in design and manufacture. But does this common-sense observation establish that thinking computers will exist within fifteen years as Kurzweil has claimed? In order to make this leap, Kurzweil must read far more significance into such historical trends, and then tie the disparate processes behind the creation of the Universe, the evolution of life, and the development of new technologies together by positing some overarching and ineluctable law of reality. And it is the universal abstraction of 'information' that makes this possible.

Kurzweil's posited law states that, over the history of a process such as evolution or the creation of new technologies, significant events happen more and more closely together, effectively 'speeding up' time, reaching effective instantaneity at the moment of the Singularity. Therefore, it took a long time for *homo erectus* to appear, but *homo sapiens* appeared in the cosmic blink of an eye after that. As shown by Moore's Law, the development of computer processing power is similarly 'speeding up'. The false reasoning here results from attributing to evolution an agency like the agency of those researchers seeking to build faster microprocessors. Evolution hasn't 'sped up' or produced more significant changes over time; it has progressed at no particular pace because it isn't going anywhere. Evolution doesn't progress in a linear fashion towards some predetermined destination and it isn't guided by an intelligence that is attempting to achieve some particular goal. Ray Kurzweil, as a human being, can scan the known archeological record and trace a line of significant evolutionary developments that led to the appearance of human beings, but of course evolutionary processes weren't trying to make human beings. Mutations only become significant in hindsight, and if Kurzweil were a dinosaur or an insect, he would identify a different set of significant changes and create a different history. A human being can write a report card for evolution (if he or she so desires), noting an initial lack of speedy progress towards the creation of human beings but commending it on applying itself more energetically to the task at a later stage, but this rather solipsistic exercise can hardly be considered objective data concerning how evolution works. The same is true for technology: human beings didn't figure out how to smelt bronze because they saw it as a necessary step towards their ultimate goal of building personal computers; it was most likely an accidental discovery, which was adopted because it served a particular use in its time. Kurzweil's teleological view of history is entirely a product of his own perspective as a human living at a particular moment, but he confuses it with a God's-eye view of how the Universe objectively functions.

Nonetheless, Kurzweil sees history as providing more-or-less conclusive proof that the Singularity awaits. Believing as he does that an age of miracles is just around the corner, Kurzweil hopes that, if he just can survive long enough, he will be able to live forever. While some might find the prospect of having their brains destroyed so that a robot doppelgänger can take over their life and personality distinctly unappealing, the reason why Kurzweil and others have greeted it with enthusiasm is, of course, a belief that this will mark the secular equivalent of the moment after the last trump, when the righteous have their dead or imperfect bodies replaced with new, perfect and immortal forms.

The eccentric Russian multi-millionaire Dmitry Itskov has even founded the '2045 Initiative', dedicated to developing the capacity to upload minds into robot avatars by its eponymous year (Segal 2013), and it is an indication of the influence of the mind uploading idea that it has twice in recent years been subjected to sustained attention from scholars. The first example is a discussion of the technological feasibility of mind uploading in *The International Journal of Machine Consciousness* in 2012, while the second was initiated by a consideration of the Singularity by David Chalmers (Chalmers 2010), who was presumably drawn to the topic because of the way in which it literalised philosophical questions with which he was already concerned.[15]

But if figures like Kurzweil and Itskov believe that they can live forever by replacing their current bodies with newly manufactured ones, what precisely about themselves does they believe will persist from one body to the next, and why do they believe that their thoughts and personalities would continue in a new body?

How could it be possible to upload your mind into a computer, and what kind of mind would it be once it got there? The mind uploading scenario hinges

---

15 Chalmers's coverage of the Singularity (which encompasses more than just mind uploading) was broadly sympathetic, although his willingness to skip over some of its problems without challenge might have arisen from a desire not to miss out on a discussion of the philosophical questions their possibility raised.

on three core beliefs: a) a computer or similar information processing device can simulate the functioning of the human brain; b) the mind of a living person can be translated into a format that would allow it to be transferred into such a device; and c) this device would then be able to generate a mind that would continue the subjective experience of that person without modification or interruption.

Given that the computational model of the brain has been dominant for some time, it is not surprising that someone like Kurzweil considers simulating the function of the human brain with a computer to be a done deal. If the brain is an information processing system that functions in more-or-less the same way as a computer, then a sufficiently powerful computer should be able to run the 'software' of the human mind. If the number of connections in the brain is used to produce an estimate of its processing power, it is currently ahead of the computer, but here Moore's Law remains the futurist's friend, stating that computer processing power will continue to grow at a predictable exponential rate until it leaves the brain behind.

As for the second of the core beliefs required for faith in the mind uploading scenario, a materialist account of the mind requires a belief that all thought, consciousness, memory, etc. is a product of neural activity in the brain and so, if sufficiently sophisticated scanning is available, it should be possible to gather and record the patterns of functioning in an individual human brain, then simulate those patterns in a computer built to reproduce its neural architecture. Having established the possibility of meeting these two engineering challenges, both Kurzweil and Moravec feel that point three will inevitably follow.

However, even if a computer could simulate my brain activity, would that computer therefore possesss a mind indistinguishable from my own? The quest for artificial intelligence more generally does not need to satisfy this third requirement: for AI to be achieved, it is necessary that a computer produce *a mind*, but it is not necessary that it produce any *particular mind*. However, the mind uploading scenario requires that machines produce, not simply some kind of consciousness, but a flawless copy of an existing, biologically based consciousness.

For the likes of Hans Moravec and Ray Kurzweil, the ability to build a conscious machine is a scientific certainty. It is a scientific certainty because the human mind is a product of the information processing power of the human brain; the only reason why we don't have conscious machines now is that we are not yet able to build machines with information processing power comparable to that of human brains. Thanks to Moore's Law, Kurzweil feels that he can even pinpoint the approximate date when machines will start to think by performing simple mathematical calculations: in 1999, Kurzweil predicted that, by 2023, a $1,000 computer will match the processing power of the human brain (Kurzweil 1999: 102–6). It might be countered that there is no particular reason to believe that, even if this comes to pass, it will produce thinking machines, but again a belief that consciousness *is* information processing suggests that it can.

However, even if this is true, it seems reasonable to expect that, if a conscious machine is one day built, that consciousness almost certainly will be utterly alien to us, so much so that it may not even be possible for human beings to understand it as consciousness at all. To think otherwise is to see consciousness as a single, objective, fixed phenomenon – this is something that any understanding of mind as product of brain function must do to a greater or lesser degree, given its belief that consciousness exists as a direct product of information processing and so should be fundamentally the same in different instances as long as information is processed in the same way in each case. It also harks back to Descartes' account of consciousness, in which human beings have consciousness, but all other living things are simply mindless machines. This suggests that consciousness is an either/or proposition: it is either absent, as in the case of non-human animals, or present, as in the case of humans, and so machines will have no consciousness until some moment when their level of processing power reaches a point when they suddenly begin having conscious experiences like those of human beings. This seems a very naïve view: there is now abundant evidence from the study of animal behaviour to establish that there are different kinds and different levels of consciousness.[16] It is also implausible that our ancestors went about their business with no consciousness up until some magical moment when they began to think and experience themselves and the world as we do. Philosopher Thomas Nagel, in a famous essay entitled 'What is It Like to Be a Bat?' (1974), argued that, given the vast differences in the variety of embodiments and sensory horizons amongst animals, we can't even imagine how other living creatures must experience consciousness.

> Our own experience provides the basic material for our imagination, whose range is therefore limited. It will not help to try to imagine that one has webbing on one's arms, which enables one to fly around at dusk and dawn catching insects in one's mouth; that one has very poor vision, and perceives the surrounding world by a system of reflected high-frequency sound signals; and that one spends the day hanging upside down by one's feet in an attic. In so far as I can imagine this (which is not very far), it tells me only what it would be like for *me* to behave as a bat behaves. But that is not the question. I want to know what it is like for a *bat* to be a bat. Yet if I try to imagine this, I am restricted to the resources of my own mind, and those resources are inadequate to the task. (Nagel 1974: 439)

If the sense of self of a bat is so alien to us, how much more so would be the mode of consciousness of a machine?

---

16   For the complexity of animal consciousness, see Panksepp (2005); for human consciousness, see Damasio (1999: 121).

Furthermore, our consciousness of our selves and world – like that of the bat – derives from our experience of embodied interaction with our environment. From the moment of birth, our experience and understanding springs from our movement and interaction with our environment and other people, from our desires and directed movements. Would a thinking machine be able to understand the metaphorical underpinnings of what we say, when concepts such as – to take one example from amongst countless others – up/down, high/low, or rise/fall are meaningless because it has no inner ear with which to sense movement, cannot stand up/sit down/lie down or fall over and does not feel the pull of gravity?[17] Researchers in behavioural robotics and Artificial Life understand that intelligence and consciousness arise from being in the world, rather than self-reflection, and so create robots and automata with some purpose and motivation, but this only illustrates the point more clearly: if a robot made to scuttle around on the floor like a cockroach developed self-consciousness – or, by analogy, if cockroaches themselves one day miraculously developed consciousness – would we really expect that consciousness to be the same as our own? These more recent approaches to creating artificial intelligence are founded on a realisation of the naïvety of those conceptions of mind that inform the mind uploading scenario. Often human beings find themselves unable to understand the thinking of other human beings who live in different circumstances; can we really expect that the consciousness of a machine will be indistinguishable from our own?

It might be objected that this is all a matter of how extensive and accurate the simulation of the human mind is, rather than the underlying nature of artificial self-consciousness itself. A computer could understand concepts like high and low if it were equipped with a sensor that registered altitude and/or gravity; all we need to do is reproduce all the inputs that feed into a human brain, and the result would be consciousness more or less identical to our own. This brings us back to the likelihood of creating a machine that can not only think, or even think like a human being, but also whose thoughts and consciousness are *identical* to those of a particular, individual human being such that they are considered to *be* that particular, individual human being.

Can Ray Kurzweil expect to still be Ray Kurzweil after undergoing a 'brain porting' procedure? After all, as I've already noted, no physical component of Kurzweil's body will continue to live after this procedure. To believe that the new entity created would be Kurzweil requires a redeployment of the Turing Test; just as a machine that can convince us it's the same as a person should be credited with being a person, so a machine that can convince us it's the same as the (recently deceased) Ray Kurzweil should be credited with being Ray Kurzweil.

---

17   See Lakoff & Johnson (1980).

The mind uploading scenario certainly entails death, even though its description obscures this. Uploading our minds, we are told, will allow us to become immortal; this is the great prize it promises, which makes people so desperate for it to be true. But think again about what precisely is being described. If Ray Kurzweil's consciousness is 'uploaded' into a computer, does that really mean he will not die? In fact, Moravec's account of how this would happen specifies that the procedure is itself fatal; the unmentioned postscript to his description would be the wheeling from the operating room of a dead, decerebrated body, which would then be buried, cremated or disposed of in some other way. To believe that life continues in this scenario, one must believe in an equivalence between body and machine that is more direct and unambiguous than any of the many we have looked at previously; it is necessary to believe that the life and mind of a real, individual body has been transformed into a machine while retaining all of its essential characteristics.

As is so often the case, the positing of a direct equivalence between body and machine depends upon a literalisation of metaphor; the words 'uploading' and 'consciousness' serve to paper over an illogicality that would be immediately apparent if different terms were used. Neither 'uploading' nor 'consciousness' can be taken at face value as descriptions of what is being proposed. Consider in detail the use of the word 'uploading'. 'Uploading', like almost all words pertaining to electronic communication and the flow of information, depends fundamentally upon analogy. Before the advent of the telegraph, all communication depended upon physical movement – to relay information to another person required either physically placing yourself in proximity to that person, or some mechanism that physically moved an artefact such as a letter into the addressee's possession. However, with the advent of electric, and then electronic, communication, this language of physical movement became metaphorical in its reference to movements and agencies that occurred outside human perception – the movement of radio waves, or electrical impulses through wires, for example. The term 'upload' suggests a direction of physical movement ('up') and the shifting of a physical presence (to load, as cargo is loaded onto or off a ship, for example). However, of course, no such physical movement of a physical artefact takes place. If I download a file from the Internet, no actual physical movement has taken place; at the end of the process, the file remains at its point of origin, and has not 'gone' anywhere. What has happened is that a copy of this file has been created on the hard drive of my computer; in other words, the file has not been moved or transferred, it has simply been duplicated elsewhere. As philosopher Patrick D. Hopkins has pointed out, the term 'mind uploading' relies on a set of metaphors to suggest that, first,

> [t]he mind is located 'in' a particular place – some brain. Using technology, we will able to move ('transfer') the mind from 'within' the brain it is currently located to another location – 'onto' another substrate or 'into' a computer or

to another 'receptacle'. Second, thinking of the mind as 'in' the brain suggests thinking of the mind in terms of substance – the mind is being treated as a thing, an object, something that is locatable and takes up specific space and that can therefore be moved from 'inside' one thing to 'inside' another. (Hopkins 2012: 231–2)

All of this is, of course, not literally true. To 'upload' my consciousness, therefore, is not to carry out some dream of transmigration, or to imbue me with immortality; rather, it means to create a duplicate or simulation of myself that can continue to run after my (still inevitable) death. Nothing would be physically transferred from Ray Kurzweil's body to an android; the only thing removed from it would be his atomised brain, which would be discarded. Kurzweil will have died – will have, indeed, been *killed* – and the android now left walking around claiming to be him would have no physical connection to him at all. Instead, the computer brain of the android would be programmed to mimic the neural firings of Kurzweil's destroyed brain. In light of this, it seems a bit of a stretch to say that this procedure has rendered Kurzweil immortal; it seems far more accurate to say that Kurzweil has died and been replaced by a computer simulation based around readings of his brain function.

This might be dismissed as mere quibbling from someone who hasn't gotten past a primal attachment to the obsolete 'wetware' or 'meat' of the human body. But the fact remains that neither Ray Kurzweil nor anyone else alive today can actually identify what the consciousness that will supposedly be transferred in this procedure actually is or how it functions. If I download a PDF file, I can read that file, show it to someone else, or print it off as a hard copy. It isn't possible to show someone else my consciousness, or translate it into some other medium or code so that I can pass it to someone else. There isn't even a universally accepted theory of what consciousness is or how it works, let alone some – even hypothetical – method of translating it into some other code (like binary code, for example) that would allow it to be 'moved' from one place to another. So how can anyone be planning to simulate it? According to a traditional materialist account of the mind, consciousness is simply an epiphenomenon of the brain; therefore the brain can be treated like a 'black box'. It isn't necessary to understand how the brain works; all that is necessary is to build a perfect replica of it. If our brains produce consciousness, then a perfect replica of our brains should also produce consciousness. But because of this, when someone speaks of uploading consciousness, they are in fact not talking about consciousness. In fact, consciousness does not directly figure in what they're talking about at all. Rather, they are talking about simulating the functioning of the human brain, and tying this to a hope that, if you simulated someone's brain perfectly, it would produce a duplicate of their consciousness as well.

The Brain in the Vat

In his original mind uploading description, Moravec argues that our patterns of neural activity alone make us who we are – in other words, we exist ultimately as information processing, and so live on for as long as that information processing continues to take place somewhere.

> Body-identity assumes that a person is defined by the stuff of which a human body is made. Only by maintaining continuity of body stuff can we preserve an individual person. Pattern-identity, conversely, defines the essence of a person, say myself, as the *pattern* and the *process* going on in my head and body, not the machinery supporting that process. If the process is preserved, I am preserved. The rest is mere jelly. (Moravec 1988: 116–17)

But *is* the rest really 'mere jelly'? The mind uploading scenario is dependent upon a belief that the mind is an immaterial phenomenon produced by the information-processing going on inside the brain. It hinges on a belief that all human thought and experience is generated by the brain, for which the rest of the body is nothing more than a vehicle and life-support system, and yet even the brain itself is only incidentally important; the mind has no materiality at all, but only requires a physical system to produce it, and evolution just happens to have provided the brain rather than David Chalmers's silicon or pipes.

Moravec's belief in the feasibility of mind uploading takes as its point of departure ideas arising within the philosophy of mind (see Moravec 1999: 169–70), but treats these as engineering challenges rather than thought experiments. For example, Chalmers has argued that it would be possible to progressively replace individual neurons in the brain with silicon chips without there being any point at which the brain stopped being a brain or human thought was interrupted (Chalmers 2003: 239ff.), and thus that there is nothing special about the relationship between the physical matter of the brain and thought. This idea would seem to establish the viability of the kind of brain scanning and simulation Moravec describes, as well as the continuity of an individual's thoughts and identities throughout, although neuroscientist Susan Greenfield has pointed out that Chalmers's conclusions have no basis in medical fact:

> No simple, systematic substitution of one silicon component after another could ever have the same effect, unless of course that unit was a simulacrum of the neuron, replete with all the chemicals and biochemical machinery that makes its characteristic, restless plasticity possible along with its variable sensitivity to whatever modulating chemicals might happen to have been released at a particular time. Whilst the hypothetical scenario of neuron substitution is conceptually logical and plausible, in reality it's meaningless and unhelpful. (Greenfield 2012: 116–17)

More famous is the 'brain in a vat' scenario first introduced (to the best of my knowledge) by Daniel Dennett in 1978 but revisited in various forms by Dennett and others since (e.g. Dennett 1981).[18] The various versions of the brain-in-a-vat scenario centre on the idea that, if your brain were removed from your body but stored in a vat capable of keeping it alive, and connected to computer equipment that sent and received electrical impulses identical to those previously sent and received by the remainder of your body, your subjective experience would continue unchanged. The intended moral of this story, obviously, is that the information processing carried out by your brain is really 'all that counts' and, whether it's in a vat or not, your sense of having a body and being located in an environment are illusions created by your brain, a kind of 'virtual reality' you experience in the theatre of the mind. This idea, which has been given its popular expression in the *Matrix* films, exemplifies the kind of implausible oversimplifications the computationalist approach can draw its proponents into.

Of course, on one level, the brain-in-a-vat scenario must be true, but it is a level so basic as to be meaningless. Yes, if my brain was removed and wired up to a system that perfectly simulated its interactions with the rest of my body, then things wouldn't be any different for me, but you could just as easily argue for a 'thumb-in-a-vat' scenario, in which my thumb is placed in a vat and wired to a system that simulated its interactions with the rest of my body perfectly. On the contrary, the brain-in-a-vat scenario is meant to establish that the brain has a status that sets it apart from my thumb – not simply that it is more important than my thumb, but that its role in who I am is of an entirely different order. In other words, it is the physical locus of my mind and personality, such that every other part of my body is incidental and expendable.

According to Dennett's fictionalised description of what it would be like to become a brain in a vat, there is no change in his experience except that he can't get drunk; his fictionalised narrative even describes feeling panic and nausea despite his having no stomach or other body parts (Dennett 1981: 222–4). But would it really be possible to experience panic without a body? Based on clinical research, the work of neurologist Antonio Damasio unequivocally indicates otherwise, leading him to state that

(1) The human brain and the rest of the body constitute an indissociable organism, integrated by means of mutually interactive biochemical and neural regulatory circuits (including endocrine, immune, and autonomic neural components); (2) The organism interacts with the environment as an ensemble: the interaction is neither of the body alone nor of the brain alone; (3) The physiological operations that we call mind are derived from the structural and

---

18   For criticisms of the brain in a vat, see Sheets-Johnstone (1999: 421–8) and Gallagher (2005: 134).

functional ensemble rather than from the brain alone: mental phenomena can be fully understood only in the context of an organism's interacting in an environment. (Damasio 2000: xvi–xvii)

Even if we limit ourselves to the business of how we think or an abstracted sense of our individual personalities, much of this is still the product of chemical processes in our bodies, of the release of hormones into our bloodstreams or the ingestion of mind or mood-altering substances. These things do not take place within the confines of the brain, even though they have an impact on how the brain functions.

However, like the computer that can't understand the difference between high and low, it might be answered that Ray Kurzweil's new robot body simply needs systems that simulate such phenomena. It might have buttons on its wrist marked 'coffee', 'cigarette', 'alcohol', or 'nap', and pressing each one would modulate its brain function to simulate the impact of these things on the individual's experiences and state of mind. Or perhaps this all sounds a bit sterile, so the android is equipped with an artificial 'stomach' that can analyse actual substances inserted into it and trigger an appropriate experiential response. Certainly, given Damasio's work on the role of the body in emotion and emotion in cognition (Damasio 2000), Ray Kurzweil's new body would need to simulate the natural changes in our blood chemistry that are happening all the time.

But the whole concept underlying the mind uploading scenario is already starting to fall apart at this point. We are much more in the realm of simulation than immortality when we no longer just have the transmigration of an immaterial mind and instead start talking about computer algorithms for simulating the effects of alcohol on a non-existent human biology. Furthermore, it becomes clear that the processes going on inside a human brain are simply not enough to plausibly recreate a dead person. If I broke my leg in an accident as a child and to this day walk with a limp and feel pain on cold mornings, which makes me more irritable than on warm ones, then my android replacement will not only need to walk with a limp, but it will need to feel simulated pain and have its personality skewed towards irritability on cold mornings in order to be a perfect replacement for me. The more we follow such examples and speculations, the more the reproduction of my neural processes recedes into the background, going from being the only consideration to just one amongst an absurdly elaborate system of simulations and artificial ruses meant to add up to a perfect recreation of an absent living body. And how does Moravec's robot surgeon decide which personal ticks, idiosyncrasies and physical peculiarities are definitional to who you are, and which are merely incidental?

But perhaps we should consider the process as closer to the mythological scenario to which it obviously owes so much: the moment when, according to

Christian belief, the last trump blows and the righteous departed are restored to life in newly minted and perfect bodies. Why would I want my new body to have a limp and give me pain? Why would I want the face that I have now, rather than one with movie-star good looks? Perhaps such complaints are pointless because the whole appeal of this process is an escape from our biological frailties and failings. This is probably true, but it is equally true that, without my previous face or limp (if I had one), I would not be the same person. If I'd never had them, my life might have been different in all sorts of ways, and my personality and personal outlook too. Without them, the android would not be *me*, but only a partial simulation of me, which even someone with whom I had only a passing acquaintance would recognise as unconvincing. Logically, to perfectly replace my dead body it would be necessary to simulate – perfectly – every aspect and attribute of the body I have now; to do anything less would be only to create a partial simulation. However, if it was possible to create such a perfect replacement, what we would have would simply be … *my current body*. It would respond to a cup of coffee or a pain in the leg in the same way that I do now because, rather than being a superhuman robot with a simulation of my neural processes running in its CPU, it would be, for all intents and purposes, another body just like the one I already have. It would then more properly be a replacement, rather than a simulation, but in that case the question becomes quite different. Rather than discussing the possibility of uploading our minds into a computer so as to live forever – an idea rendered nonsensical and self-evidently impossible by such a line of thought – it becomes a question of whether science could ever reconstruct – molecule by molecule, presumably – a perfect replica of an individual human body. Even if an attempt to maintain the mind uploading scenario was based on the idea that clever computer algorithms could simulate all the processes of the human body in some more economical way, this still establishes that Kurzweil's timeline for the arrival of immortality is flawed. Even if future increases in computing power were following some immutable physical law rather than being simply products of human industry and therefore beyond prediction, Kurzweil's predictions are fixed on a posited moment when processing power will match the processing power of the human brain. Once it is realised that human experience and consciousness are not simply a result of the processing power of the human brain, but depend upon a vast array of interrelationships both within the whole human organism and between the organism and its environment (see Johnson 2007: 1–2), the attainment of this level of computing power no longer becomes the landmark it has been claimed to be. Rather than being the moment when it will be possible to simulate human consciousness, it becomes the moment when it is – perhaps and hypothetically – possible to simulate one component of human consciousness. The brain is the most complex component of human consciousness, certainly, but this does not mean that simulating it would create an effective reproduction of who you are along the lines of the mind uploading scenario.

So even if the patterns of information in Ray Kurzweil's brain can one day be recorded and reproduced on a sufficiently sophisticated computer (and assuming his health regime keeps him alive long enough to undergo the procedure), Kurzweil's assumption that everything that makes him who he is exists as a kind of computer program running in his brain is misguided. Much of his personal experience, and even his capacity for normal human cognition, would be lost. Unless his new artificial body is carefully constructed so as to have the same weight distribution, the same amount of muscle strength, and the exact same proportions, he presumably would not even be able to walk or pick up an object until he had relearnt these basic motor skills, and so he would be left lying on his back in an oversized playpen, cognitively and physically crippled.

But a boundless faith in the ability of new technologies to solve any problem can neutralise even this threat to the mind uploading scenario. Some time long after Kurzweil's fifty year horizon for immortality, he speculates that advanced nanotechnology will allow the creation, artificial cell by artificial cell, of a perfect replica of his body, which will then presumably have all the embodied capacities of his current, biological body. Of course, believing as he does that everything important is going on inside his brain, to him this kind of body seems more a luxury than a necessity, something he can upgrade his interim artificial body to when it becomes available. Kurzweil also takes on the speculations of nanotechnology researcher J. Storrs Hall regarding 'utility fog'. Utility fog is a swarm of microscopic 'foglets', tiny robots that can hover invisibly in the air, linking together to create a physical presence – such as a computer or an armchair, for example – as needed (see Hall 1993, 2001). Kurzweil imagines his consciousness one day existing as a pattern of information in a network, able to instantaneously create itself a body out of utility fog whenever a physical presence becomes necessary before disappearing back into the immaterial realm of information (see Hall 2001).

Kurzweil's utility-fog body represents a belief in the possibility of an ultimate victory of informatic and computational models of mind over embodiment. Nanotechnology, as a technology aimed at creating entirely new matter according to human specifications, represents the ultimate reduction of matter to information, creating a world where physical forms are a product of information systems. The utility-fog body reflects a belief that the body is nothing more than extension, a dispensable tool in the service of an immaterial mind. However, it also brings a concern with the material back into the equation, returning the mind to a physical form – even if only temporarily. Predictions about a coming age of nanotechnology – which figure in the singularitarian epiphanies of the likes of Ray Kurzweil – represent both a final triumph of the informatic view and the colonisation of all matter by information, but also unavoidably return to a preoccupation with matter. It is to this shift that we will turn in the following chapter.

144

Clearly, not every – and probably not even most – adherents to the idea that the mind is produced entirely by information processing in the brain believe that their minds could one day be instantiated in computers. At the same time, this idea does little more than literalise or alter in magnitude but not quality claims made by philosophers of mind regarding computationalism. From the middle of the twentieth century, the computer – immediately characterised as a mechanical brain – became the dominant machine analogy for the human body, but in doing so it shifted the nature of this perceived relationship in fundamental ways. Where previously the commonality of machine and body was based on their shared capacity for physical movement and the idea that both were pieces of physical engineering, this new approach sees, not movement or energy, but immaterial *information flows* as definitional to bodily life. Bodies are products of information storage and processing going on in DNA, and are really little more than tools used by the brain, which is itself an information processing system that produces the mind. The appearance of this understanding did not challenge the older idea that the body is a machine, but it added to it a mechanistic account of what had previously been the stubborn remainder to mechanistic accounts of the body: the soul or mind. The mind remains immaterial and divorced from the body, but the idea of information as an immaterial, universal substance allows the mind to also be accounted for in mechanistic terms. Immaterial information systems lie both below the bodily mechanism in the DNA, and above the bodily mechanism in the mind; the mechanisms of the body are simply to some degree arbitrary physical systems necessary to get from one to the other. Not only does this mean that the body is simply a tool of the computer mind that might safely be discarded, but also that even the brain ultimately can be dispensed with too. AI research is founded on this idea: artificial minds can be created simply by reproducing the patterns of information processing going on in the human brain with a machine, then providing that machine with inputs and outputs that replace the body. Taking this idea to its logical conclusion, it should be possible to dispense with the messy, inefficient and failure-prone mechanism of the brain and replace it with something else entirely that can process information in the same way. As long as the new mechanism can perform the same information processing functions, the immaterial true self of the mind can live on.

Chapter 5
An Aesthetics of the Invisible[1]

The Matter of Invisibility

In the last chapter, we saw how the discovery of DNA allowed molecular biology to tie the composition of the human body into the framework of information theory. This was a key factor in the development of a computationalist explanation of the body – that is, the application of a kind of mechanism that depended specifically on the computer as its machine model. DNA has come to be the most famous and recognisable of all molecules; with its distinctive two helices like the serpents of the caduceus, it is now well established as a kind master sign, the shape that most fundamentally encapsulates life on earth.

For this reason it can produce a kind of mental jolt to remember that no-one has ever actually seen DNA and, furthermore, no-one ever will. The DNA molecule, and every other kind of molecule that forms the structures of our bodies and material reality, have no visual attributes. Although each of us has seen DNA countless times, it has no appearance. It is, simply put, invisible.

Molecules are therefore perfectly situated to serve as the material anchors for explanations based on the invisible, immaterial stuff of information. DNA cannot be seen, and yet it is believed to determine the shape and nature of the gross physical stuff of the human body. It seems to mediate between the formless, invisible realm of information and the physical world accessible to our senses.

When I say that DNA is invisible, it is important to understand that this is not simply a matter of DNA being too small to see with the naked eye; it *is* about how small it is, but it's more complicated than that. It's not just about needing a more powerful microscope: DNA is smaller than the wavelength of light, and as a result it doesn't interact with light, and so can't generate any visual information (Goodsell 2006: 44). Vision works by capturing light reflecting off objects in our environment, but light doesn't reflect off DNA – therefore DNA is invisible. Even if you were shrunk down to the size of DNA, incredible shrinking man-style,[2] you still wouldn't be able to see it. In fact, you wouldn't be able to see anything, as your eyes would have become much too small to interact with light themselves. You'd be blind.

---

1 Material from this chapter has previously appeared in a slightly different form in Black, D. (2014), 'An aesthetics of the invisible', *Theory, Culture & Society*, 31(1), pp. 99–121.

2 Ignoring for the sake of the argument the physical impossibility of shrinking all the molecules in your body to such a size.

The double helix structure of DNA wasn't discovered because someone saw DNA. It resulted from a process by which various people attempted to formulate a way of representing the known physical properties of DNA. Watson and Crick famously succeeded in producing a model of DNA using the double helix structure in 1953, but they did so largely by trial and error, not the use of visual clues (Kemp 2006: 64–7). Although the X-ray diffraction pattern of Rosalind Franklin's 'photograph 51' was clearly a crucial reference for them, this was not a photograph of DNA as such, but only a visual representation of certain of its qualities. The DNA molecule has only ever appeared to us through diagrams and visualisations, and today these visualisations are predominantly created by computers.

Watson and Crick's model of DNA was, literally and materially, a model. They made pieces that represented what had to go into the model, and then tinkered with them until they could get all the components into the same structure. DNA then became visualised as a rather jerry-built assemblage cobbled together largely out of bits and bobs from around a laboratory. Nowadays we usually see it as slick computer graphics, which are naturalised by a wider blurring of the boundaries between physical reality and CGI, but even these are just artefacts that follow certain conventions and styles that arise from human technologies and sensibilities. We're not objectively seeing DNA, as this simply is not possible; we are rather consuming a historically and culturally specific mode of representing it. Writing of Watson and Crick's famous model, the art historian Martin Kemp draws attention to the fact that this can only be a human representation with its own distinctive style:

> Looking at the now famous photograph of the young American and English scientists posed with their creation – taken by Barrington Brown for *Time* magazine but not published until later – it is difficult not to be struck with how much both the model and the photograph (with haircuts and suits) are 'of their period' – that is to say that they breathe a style which anchors them in time. Our sense of the period style of the Watson–Crick model – linear, wiry, openly mechanical, unadorned, and rhetorically 'functional' – is necessarily framed by reference to earlier and later systems of representation. Their precarious spatial lattice stands very much within the design parameters of the 1951 Festival of Britain, the event which decisively marked the pragmatically British embrace of a modern style for a new age. In contrast, the 'Glyptic Formula' kits for modelling molecules in the nineteenth century, with their polished balls, firm rods, and turned mahogany stands, exude the air of a gentleman's billiard room; while recent computer images ostentatiously parade the high-tech rhetoric of electronic graphics … As in any work of design, more is involved than structure and function. Whether we call the visual ingredient 'style' or 'aesthetic', we can intuitively sense its presence – in both conventional and ground breaking artefacts. (Kemp 2006: 66–7)

Any representation of a molecule relies on a set of conventions regarding what kind of information about the molecule should be communicated. The best-known convention is the 'Tinker Toy' style of 'balls and struts' diagrams, but this is only because the connections between atoms in molecules is most often the key information the diagram is trying to capture. Furthermore, the fact that contemporary, computer-generated representations of molecules also take the balls and struts form results from the inheritance of a set of representational conventions by a representational technology that is no longer subject to them. In this case, a particular style and medium of representation (the physical ball and strut model-making components) has become naturalised so successfully that it is no longer seen as dependent upon the historical or material circumstances that produced it.

This is not to say that this is the only representational convention for molecules. There are others – such as orbital diagrams or the ribbon diagrams used to capture the folding of proteins – and computer imaging allows for a greater variety of modes of representation, but this variety only draws attention to the fact that there is no single, objective or unproblematically truthful visual representation of a molecule. None is objectively more accurate than another because, as I've already said, molecules don't objectively look like anything. A representation that aimed to approximate most closely the physical qualities of a molecule would look like a squashed blob, as there aren't really any elegant struts of the kind seen in the 'Tinker Toy' type diagram holding the atoms together.

The representational conventions introduced by human researchers are designed to aid human understanding and even the manipulation of molecules by artificially integrating them into the human sensorium. While the realm of molecules has been a source of fascination since the nineteenth century, the more recent appearance of machines able to intervene at the molecular scale has given rise to the hope of eventual human mastery over this invisible domain. The much-hyped and anticipated development of nanotechnology is taking place at this invisible scale, but this research does not simply seek to delineate its structures – it seeks to assemble new ones using new machines designed to make the invisible visible.

The nanotechnological endeavour and its expected outcomes cannot be fully understood without situating them within the historical development of ideas covered in the preceding chapters. The diagrammatic, abstract, informatic understanding of life and action, and its origins in a long-running tradition of seeing materiality as a kind of incidental dead weight, is foundational to the perceived purity and power of the molecular realm, and promises a comprehensive dissolution of distinctions between living bodies and machines. After providing an introductory overview of nanotechnology and how it has been presented, therefore, I will move on to a discussion of the ways in which it redeploys much older ideas about the geometric perfection of atoms and molecules, the universality and generative power of information, and the mechanistic principles that underly life.

Nanotechnology

Nanotechnology is a technology that, by definition, exists at the molecular scale, or nanoscale. A nanometre is one millionth of a millimetre, and nanotechnology is generally understood to exist at scales of less than 100 nanometres (Freitas 2005: 243). The wavelength of light is around 400–700 nanometres; in other words, nanotechnology exists entirely in the invisible realm of the molecule.

The term nanotechnology was popularised by the 'father of nanotechnology', K. Eric Drexler, who wrote a PhD thesis on the idea at MIT in the 1980s. During the process, Drexler wrote an influential book called *Engines of Creation* (1990), which captured the imagination of a popular audience by describing nanotechnology in terms of such sci-fi stalwarts as immortality, the creation of artificial intelligence and the colonisation of space. However, Drexler wasn't actually the first person to use the term nanotechnology, and it quickly came to be applied more broadly than he'd intended, to any kind of work or fabrication on the nanoscale. Drexler's proposal specifically concerned what he went on to call 'molecular manufacturing' or 'molecular nanotechnology' in an effort to differentiate it (Phoenix & Drexler 2004: 870). It is this particular conceptualisation of nanotechnology that has had the greatest popular impact by promising a future in which existing methods of manufacturing and primary production are obsolete, and the global economy is radically transformed. Governments and private companies are already jockeying for an advantageous position in the expected new order: the largest undertaking is the National Nanotechnology Initiative in the United States, which began in 2000 and has already received nearly US$18 billion in funding (Anon. 2013b), but there are equivalent initiatives in Europe, Japan and elsewhere.

Robert Freitas, Jnr., the first person to be granted a 'mechanosynthesis' patent,[3] offered some conjecture on molecular manufacturing's future impact in a 2006 article. In a future where individuals could manufacture their own goods, perhaps 'personal nanofactories' (PNs) would be given away to consumers on a plan, like a mobile telephone, and providers would generate profits through the provision of the blueprints that the nanofactory would need in order to churn out a new pair of shoes or computer (Freitas 2006):

> Our preliminary analysis begins with an assumption that at the end of a 20-year period of introduction, almost every household in a given developed country has purchased a PN. The PN will be capable of building any manner of

3   That is, a patent for a (still only hypothetical) molecular manufacturing tool (Peterson 2010).

consumer goods using simple molecular feedstock such as acetylene or propane gas that will be piped into the home via a utility connection, similar to present-day hookups that deliver natural gas, water, and electricity …

If we further assume that (1) the price to acquire the PN is approximately US$4400 …, (2) the PN has a mass of 10 kg and produces consumer products at the rate of 1 kg/h, and (3) the PN is operated 50% of the time throughout a useful lifetime of 10 years, then the PN during its useful life produces 44,000 kg of consumer products which then have an amortized capital cost of $0.10/kg, a cost that is built into every product manufactured by the PN …

The PN will be a versatile appliance, able to manufacture whatever is deemed legal in the 21st century such as: (1) consumer goods including nondurables such as food and durables heavily laden with nanosensors, nanomotors, nanopumps and nanocomputers, (2) all the patterned sheets or chunks of diamond that anyone might want, and (3) some kinds of medical nanorobots for personal use, though these may be heavily regulated. We assume that the PN will not be allowed to manufacture contraband, nor various types of weapons systems including ecophages, or more PNs (which would nullify the R&D funding and manufacturing business model). If the public is not allowed to manufacture PNs using PNs, then the production cost of a PN using a PN becomes almost irrelevant to the retail price of a PN. The manufacturer may charge $4400 for a PN even if it only costs them $10 to make one …, using the difference to pay for showrooms and sales staff, marketing and advertising, legal costs for defending the brand and a monopolistic pricing regime, product liability insurance costs (which could be substantial), warranty and servicing costs, online and print publications including consumer how-to books and magazines, websites and online help desks, executive overhead and corporate perks, and perhaps modest dividends from profits for the shareholders. (Freitas 2006: 1–3)

The miraculous potential attributed to nanotechnology results from the fact that *everything* in our physical environment, including ourselves, is ultimately made from various combinations of around one hundred different kinds of atoms, arranged into molecules. Seeing matter as composed of patterns of atoms and molecules provides a material complement to the pervasive idea of information as the underlying structure and forming power of reality; while this view of information depends upon the abstraction of reality into a code through the unacknowledged medium of metaphor, reduction to the molecular scale provides a view of reality in which the richness of matter resolves into rearrangeable patterns reminiscent of information. Atoms exist as a relatively small collection of basic building blocks underlying the physical world; it is only the combinations and patterns of atoms that produce the diversity of everyday experiential reality. Just like the geneticist's dream of reducing the disconcerting variety and unpredictability of human diversity to different combinations of genetic code, the study of matter at the nanoscale promises the possibility of manipulating the 'DNA' of matter in general, altering and reforming it to suit human needs and fancies through the creation of new combinations of code.[4] The utopian stream of nanotechnology discourse therefore has its origins in the informatic discourses discussed in the previous chapter. Just as the human body has been reduced to patterns of information in DNA and the brain, this account expands the field until all matter, human and non-human, animate and inanimate, is reduced to informatic patterns of atoms in molecules.

The product of nanotechnological speculation that has most captured the popular imagination,[5] firing popular interest in the field and providing various heroes, villains, MacGuffins and *dei ex machina* for numerous science fiction books and films, computer games, and utopian, dystopian and extropian visions of the future, is the nanobot.[6] The nanobot is a nanoscale machine that, although it does not exist, is hoped by some to one day bring an unprecedented revolution in human life. This imagined revolution focuses upon two attributes of the nanobot. First, the nanobot is expected to give human beings the ability to 'reprogram' physical reality. That is, through the agency of the nanobot, matter can be disassembled and reassembled to order. Diamonds, infinitesimally small computers, top quality Kobe beef – the right kind of nanoscale robots (or 'assemblers' – see below) could simply cobble matter together to order out of the necessary molecular or atomic building blocks. Second, because the nanobot exists at the nanoscale and is itself built from the fundamental building blocks of matter shared by both living bodies and machines, distinctions between machine and living organism do not seem to apply to it. Nanobots will be like artificially fabricated bacteria, made to order so that they can perform useful functions. Too much pollution in the air? We could build nanobots that

---

4   In fact, in 2011 the US Office of Science and Technology Policy announced a 'Materials Genome Initiative', reflecting this logic of molecules as the 'DNA' of all matter in general (Kalil & Wadia 2011).

5   And a search for nanotechnology-related books on Amazon.com will demonstrate the level of popular interest in nanotechnology. According to Joachim Schummer, 'For many readers, who do not have a purchase record of other science and technology books, nanotechnology seems to be the first field of science and technology in which they invest a considerable interest ...' (Schummer 2005: 180), and the jumble of books containing serious science, speculation, and science fiction on the subject has caused unease amongst researchers like Schummer, who are concerned with the development of popular understandings of the field.

6   The substitution of nanobots or 'nanites' for clunky machinery in the reimagined *Bionic Woman*, mentioned in the previous chapter, is only one, relatively minor, example of this. Like nuclear radiation and genetic engineering before it, nanotechnology, as a new and unfamiliar area of technological endeavour in which relatively little actual knowledge and achievement currently exists, provides ample space for wild speculation, and is perfect for creating superheroes, super machines, and various other phenomena associated with a certain kind of genre narrative.

digest pollutants and leave harmless – or even beneficial – waste. Fatty deposits in your arteries? You only need welcome into your bloodstream a colony of nanobots, which feasts on the fat. At the molecular scale, it seems, distinctions between data and matter, body and machine, no longer apply. Following the rise of a belief that matter can be reduced to information, and that machines and bodies are fundamentally the same, it is no wonder that nanotechnology is expected to be the crowning achievement of human technological endeavour, unifying currently separate branches of science in a common project (see Baird et al. 2004: 9–73) and giving human beings God-like authority over physical reality. Of course, a belief in the possibility of creating such nanomachines depends upon a mechanistic, reductionist view of nature, and this 'atomistic reductionism' (Thacker 2004: 121) can hardly be said to have arisen simply in response to research at the nanoscale. Rather, these ideas have been imported into nanoscale research and speculation from already dominant frameworks of scientific understanding.

If nanoscale research is approached with a set of mechanist assumptions, nanoscale structures will inevitably be understood as tiny machines. Because research into molecular structures is informed by the idea that these structures are the fundamental building blocks of matter from which all biological systems are made, and that '[a]ll biological and man-made systems have the first level of organization at the nanoscale (such as a nanocrystals, nanotubes or nanobiomotors) where their fundamental properties and functions are defined' (Roco 2003: 337), it therefore follows that all living matter is machinic in nature. Older mechanist assumptions, therefore, confirm themselves.

Because the dynamism of both machines and cells is generated by nanoscale structures, at this level there is no material basis for differentiating machine from cell. As a result, a bacterium might be understood as a machine (powered by a 'nanobiomotor'), or a machine might be understood as a bacterium. Not surprisingly, the former option is far more commonly exercised than the latter.

There are two obvious reasons for this mechanisation of the nanoscale. The first, of course, is simply that mechanistic accounts of biology and nature enjoy a dominance that predates nanoscale research. Neither Descartes nor La Mettrie could have imagined nanotechnology, but they would have understood the idea that biological life is powered by mechanisms, even mechanisms vastly smaller than the hydraulic, magnetic or clockwork mechanisms of their day. While the first reason therefore results from factors independent of scale, the second reason is strongly influenced by the scale of this research. When searching for visual metaphors, it is inevitable that metaphors derived from structures at the human scale will be applied to structures at the non-human scale. Machines are objects we see and interact with in our daily lives; therefore, arguing that the bacterium is like a machine provides a familiar framework for understanding the bacterium. On the other hand, saying that a machine is like a bacterium has little appeal, given that we feel far less familiar with bacteria than machines. Even without the mechanistic description of the bacterium generated by nanotechnology, most of us would probably still understand the nature of bacteria by applying a human-scale analogy: for example that the bacterium is like a tiny version of a human-scale animal. The understanding of novel phenomena is necessarily heavily reliant on metaphor – the use of something familiar to provide a conceptual grasp of something unfamiliar – but the size of molecular structures encourages a particular reliance. As previously noted, structures at the nanoscale are not only too small to see with the naked eye, but are also too small to possess any visual attributes at all. While the anatomical diagram or the MRI scan might generate representations of the body that differ in important ways from the material body, the interior of the body nevertheless possesses at least potential visual attributes. If I cut open a cadaver, I might need to arrange and pose its interior in order to make sense of its contents, but the exposed viscera still possess a visuality independent of this. However, while researchers of both anatomy and nanoscale structures depend heavily upon visual understanding, the nanoscale structures have no visuality beyond that given to them by simulatory computer technology or human imagination. Therefore, if one expects to see a machine at the nanoscale, one will, because the visual reality of structures at the nanoscale is determine entirely by the human beings who represent them (or at least build the machines that do).

Molecular Machines

Being able to build molecular machines would be very handy – we could automatically remove all the excess carbon dioxide from Earth's atmosphere, have houses that paint and repair themselves and clothes that change colour made from threads stronger than steel, nearly indestructible cars made from diamond and nanobots inside our bodies that destroy infections and cancer cells, as well as all manner of other wonders. However, nobody has come close to building such a thing yet. Because the term nanotechnology is generally used to refer to both the idea of molecular manufacturing, the creation of self-organising structures at the nanoscale, and even sometimes the conventional creation of artefacts on a very small scale (like microprocessors, for example), it is hard for most people to differentiate between what might be possible and what's already happened when they come across the often breathless popular reporting of the area. In as much as a popular audience is aware of nanotechnology, reports of its use in the creation of cosmetics and other everyday materials can be tied together with science fiction scenarios of an endlessly multiplying army of robot bacteria turning the world into a blob of 'grey goo', but in reality these things are very different.

Fig. 5.1   A C$_{540}$ 'buckyball' fullerene. Brian0918 / Wikimedia Commons.

Artefacts *are* being produced at the nanoscale now, the most famous nanostructures being fullerenes, which are geometric arrangements of carbon atoms (Figure 5.1). However, these structures are self-organising – that is, they appear when a materials chemist puts the right atoms together in the right environmental conditions (heat, pressure, medium etc.) to make them spontaneously assume a certain form. Another way to produce nanostructures is to make them out of proteins, which fold up into various shapes by themselves. But the key point is that, while these things can be very useful, they haven't been created using molecular manufacturing – the exact arrangement of atoms or joining of smaller molecular building blocks. These things build themselves as long as they're in the right conditions; in contrast, Drexler's idea is to build little machines and factories at the nanoscale and then in turn use them to build whatever further materials we want.

To do this would require the ability to build things in much the same way that they're built in normal factories, but on a vastly smaller scale. The basic requirement would be a nanomachine called a molecular assembler:

These second-generation nanomachines … will do all that proteins can do, and more. In particular, some will serve as improved devices for assembling molecular structures. Able to tolerate acid or vacuum, freezing or baking, depending on design, enzyme-like second-generation machines will be able to use as 'tools' almost any of the reactive molecules used by chemists – but they will wield them with the precision of programmed machines. They will be able to bond atoms together in virtually any stable pattern, adding a few at a time to the surface of a workpiece until a complex structure is complete. Think of such nanomachines as assemblers.

Because assemblers will let us place atoms in almost any reasonable arrangement, they will let us build almost anything that the laws of nature allow to exist. In particular, they will let us build almost anything we can design – including more assemblers. The consequences of this will be profound, because our crude tools have let us explore only a small part of the range of possibilities that natural law permits. Assemblers will open a world of new technologies.

Advances in the technologies of medicine, space, computation, and production – and warfare – all depend on our ability to arrange atoms. With assemblers, we will be able to remake our world or destroy it. (Drexler 1990: 14)

However, the molecular assembler remains only a theoretical possibility. Various tiny components, dubbed 'gears', 'axles', 'hinges' or 'springs' by their creators, have been created in the lab, but certainly no-one has built a functioning nanomachine to date (Coskun 2012: 19–20). In fact, there is disagreement – sometimes bitter disagreement – amongst some nanotechnologists concerning the very feasibility of such an endeavour. Because nanotechnology is a new area, it has brought together researchers from different disciplines, and there is something of a division between the perspectives of those coming from information technology and engineering, on one hand, and those coming from chemistry and the material sciences, on the other (Bensaude-Vincent 2007: 227ff., Schummer 2004). This division was personified by the debate between K. Eric Drexler and the late Richard Smalley, a professor of chemistry and physics who shared a Nobel Prize in 1996 for the discovery of fullerenes, and who maintained that nanobots 'will never become more than a futurist's daydream' (Smalley 2001: 76).

Smalley repeatedly and relentlessly attacked the idea of molecular manufacturing on multiple grounds that all basically hinge on the fact that you can't just pick atoms up and stick them where you want them to go (Smalley 2001, Smalley et al. 2003, Bueno 2006, see also Kurzweil 2005: 236–41). They have their own intrinsic qualities, which make some stick together and some not, and will make them assume some shapes and not others. This criticism is well illustrated by the famous IBM 'Big Blue' logo that was created by Don Eigler in 1989 by individually positioning 35 xenon atoms (Figure 5.2). While this was hailed as proof that human beings can now manipulate individual

Fig. 5.2   Don Eigler's 1986 IBM logo written in xenon atoms. © IBM Corporation. Image reproduced with permission.

atoms in precise ways, it was in reality a very limited demonstration. It was only a two-dimensional arrangements of atoms (building nanomachines would require their arrangement in three dimensions), it used xenon, which – as an inert gas – is relatively easy to control, and it didn't actually bond anything together. The xenon atoms were just sitting on the atoms below, kept stable by producing the whole thing in a super-refrigerated state. Any disturbance and the entire assemblage would have flown apart (see Hessenbruch 2004: 140–41).

This division within nanotechnology research brings to mind the struggle between Gestalt theory and molecular biology at the dawn of the era of genetics, in which the reductionist, informatic approach of the molecular biologists vanquished an approach that highlighted emergent, system-level phenomena. Bernadette Bensaude-Vincent draws a contrast between 'reductionism' and 'emergentism', deriving from 'mechanistic' and 'organicist' paradigms respectively, in nanotechnology (Bensaude-Vincent 2007: 232). For the materials chemists, nanostructures are emergent systems to be coaxed into organising themselves, while for the engineers atoms and molecules are inert matter awaiting the imposition of informatic form.

Unfortunately, the emergentist understanding of nanoscale phenomena does not capture the popular imagination in the way that the tiny robots of Drexler's 'Meccano set' (Bensaude-Vincent 2007: 225) nanotechnology do. The likes of Richard Smalley might be captivated by the elegance and dynamism of structures that organise themselves into something greater than the sum of their parts,[7] but the man in the street is not likely to choose this over the promise of *Fantastic Voyage*-style submarines inside our bodies and *Star Trek*-style replicators magicking up any object that satisfies our latest whim.

Smalley certainly had the advantage of actually having manufactured nanostructures, whereas nobody has yet created a nanomachine of the type

---

7   For example, Smalley uses such metaphors as 'a boy and a girl fall[ing] in love' and dancing to describe the interactions of molecules (Smalley 2001: 76).

proposed by Drexler. The Foresight Institute, a non-profit organisation founded by Drexler, offers the Richard Feynman Grand Prize,[8] which promises $250 million to the first group or individual to produce the two key components of a molecular assembler: a tiny robot arm (like the kind of robot arm you see on car assembly lines only millions of times smaller) and a very basic computer (Freitas & Merkle 2004: 125). So far, nobody has claimed the prize.

A division is therefore apparent within nanotechnology research itself, between those who see manufacturing on this scale as a special, self-organising and emergent mode of production whose laws are fundamentally different from those of the macroscale, and those who see nanotechnology as subject to mechanistic control and organisation that generally follow the laws of existing manufacturing, and who wish to 'improve' the chemists' approach by gaining direct control over atoms and molecules in order to make whatever they wish. It is perhaps not surprising that materials chemists such as Smalley have not appreciated Drexler's tendency to characterise their endeavours as inferior, transitional work that will be superseded by his promised (but so far undelivered) vision of 'nanofacture'. Drexler's claims concerning the nanotechnological future fundamentally rely on a replacement of the discourse of chemistry employed by Smalley and his ilk with the discourse of engineering and information processing.

> Enzymes and hormones can be described in mechanical terms, but their behavior is more often described in chemical terms.
>
> But other proteins serve basic mechanical functions. Some push and pull, some act as cords or struts, and parts of some molecules make excellent bearings. The machinery of muscle, for instance, has gangs of proteins that reach, grab a 'rope' (also made of protein), pull it, then reach out again for a fresh grip; whenever you move, you use these machines. Amoebas and human cells move and change shape by using fibers and rods that act as molecular muscles and bones. A reversible, variable-speed motor drives bacteria through water by turning a corkscrew-shaped propeller ...
>
> Simple molecular devices combine to form systems resembling industrial machines. In the 1930s engineers developed machine tools that cut metal under the control of a punched paper tape. A century and a half, earlier, Joseph-Marie Jacquard had built a loom that wove complex patterns under the control of a chain of punched cards. Yet over three billion years before Jacquard, cells had developed the machinery of the ribosome. Ribosomes are proof that nanomachines built of protein and RNA can be programmed to build complex molecules. (Drexler 1990: 8)

---

8 Named after the Nobel Prize-winning physicist appropriated by nanotechnologists as the originator of their project because of a famous 1959 lecture he gave called 'There's Plenty of Room at the Bottom' (Feynman 1959).

Like his admirer Ray Kurzweil, with his 'Law of Accelerating Returns', Drexler seeks to naturalise his predictions for the future by setting them within a sweeping metanarrative that renders molecular manufacturing an inevitable destiny for human – and even non-human – 'progress'. In *Engines of Creation* he references Richard Dawkins on genes and 'memes' to cast nanomachines as successors to DNA, RNA and 'natural' evolution (Drexler 1990: 21ff.).

> Our ability to arrange atoms lies at the foundation of technology. We have come far in our atom arranging, from chipping flint for arrowheads to machining aluminum for spaceships. We take pride in our technology, with our lifesaving drugs and desktop computers. Yet our spacecraft are still crude, our computers are still stupid, and the molecules in our tissues still slide into disorder, first destroying health, then life itself. For all our advances in arranging atoms, we still use primitive methods. With our present technology, we are still forced to handle atoms in unruly herds.
>
> But the laws of nature leave plenty of room for progress, and the pressures of world competition are even now pushing us forward. For better or for worse, the greatest technological breakthrough in history is still to come. (Drexler 1990: 3–4)

Molecular manufacturing is cast as continuing a progression not only from chemistry, but from the natural formation of proteins, bringing 'a revolution without parallel since the development of ribosomes, the primitive assemblers in the cell' (Drexler 1990: 21). Repeatedly, the proof of nanomachines' viability is said to lie in the existence of living cells, which are cast as simply naturally occurring nanomachines themselves. Key to Drexler's initial justification of molecular manufacturing was a chart with columns for 'Technology', 'Function', and 'Molecular example' intended to demonstrate that naturally occurring structures were, in effect, mechanical components (Drexler 1981: 52–76), and the use of the term 'molecular machines' to refer to complex biological entities which exist at the molecular scale (Drexler et al. 1991: 76) formalises the equivalence between nanomachines and living cells and proteins. Foundational to this strand of nanotechnological discourse, then, is an older literalised metaphor, that between living systems and machines. This metaphor is very old, as we have already seen, but its post-DNA evolution towards the idea of life as specifically *information processing* machinery is crucially important. Bernadette Bensaude-Vincent has already pointed out the inherent contradictions in such a view once the creator/God of Descartes' mechanism is removed.

> Over the past decades, the machine metaphor has invaded the language of biologists. In the early times of molecular biology, such metaphors were exclusively used for DNA transcription and translation. Nowadays each entity active in the cell is described as a machine: ribosomes are assembly lines, ATP

synthases are motors, polymerases are copy machines, proteases and proteosomes are bulldozers, membranes are electric fences, and so on ... Although biologists generally agree that living systems are the product of evolution rather than of design, they describe them as devices designed for specific tasks. Indeed, if biology can teach us about engineering and manufacturing, it is because the living cell is now viewed as a factory crowded with numerous bionanomachines in action. (Bensaude-Vincent 2006: 9)

This perspective is the most crucial foundation for the claims of Drexler, who promises that molecular manufacturing will allow human beings to take control of the 'technology' of life itself in order to emulate – and exceed – its wonders.

Technology-as-we-know-it is a product of industry, of manufacturing and chemical engineering. Industry-as-we-know-it takes things from nature – ore from mountains, trees from forests – and coerces them into forms that someone considers useful. Trees become lumber, then houses. Mountains become rubble, then molten iron, then steel, then cars. Sand becomes a purified gas, then silicon, then chips. And so it goes. Each process is crude, based on cutting, stirring, baking, spraying, etching, grinding, and the like.

Trees, though, are not crude: To make wood and leaves, they neither cut, grind, stir, bake, spray, etch, nor grind. Instead, they gather solar energy using molecular electronic devices, the photosynthetic reaction centers of chloroplasts. They use that energy to drive molecular machines – active devices with moving parts of precise, molecular structure – which process carbon dioxide and water into oxygen and molecular building blocks. They use other molecular machines to join these molecular building blocks to form roots, trunks, branches, twigs, solar collectors, and more molecular machinery. Every tree makes leaves, and each leaf is more sophisticated than a spacecraft, more finely patterned than the latest chip from Silicon Valley. They do all this without noise, heat, toxic fumes, or human labor, and they consume pollutants as they go. Viewed this way, trees are high technology. Chips and rockets aren't. (Drexler et al. 1991: 19)

It is difficult to imagine what working definition of 'technology' might allow a tree to be logically referred to as such, but, as with the hollowing-out of the definition of 'information', this account of nanotechnology depends upon an emptying of the terms 'technology' and most importantly 'machine' such that they no longer require design, human fabrication, or intelligent application or purpose. Although mechanism is widely associated with the atheistic stance of La Mettrie and his ilk, in its early days during the seventeenth century it was put forward as proof of the existence of God; after all, if bodies are machines, by definition there must be someone who made them (see Sawday 2007). As Georges Canghuilem has said, 'the construction of a machine can be understood neither without purpose nor without man. A machine is made by man and for man, with a view toward certain ends to be obtained, in the form of effects to be produced' (Canguilhem 2008: 86). However, just as information can become a meaningless, purposeless and endless activity engaged in by all matter, so the animation of organic life makes it a meaningless, purposeless machine, a tool with neither maker nor user.

Drexler and others claim that molecular machines must be possible because they already exist in the form of proteins, cells and so on. However, again, this argument rests on treating a metaphor as literally true. Proteins aren't machines in any meaningful way, as they haven't been engineered by some other party. the *Oxford English Dictionary* defines a machine as a 'material structure designed for a specific purpose', or an 'apparatus constructed to perform a task or for some other purpose'. Machines are structures intentionally created in order to serve some particular use, but the attribution of machines to nature without a Cartesian belief in a divine creator suggests the existence of machines with neither use nor creator. It's possible to call cows machines for producing meat, or the Earth's atmosphere a machine for turning windmills, but to do so is to act as if they have been made to order so as to carry out certain tasks for human beings. Any system will interact with its environment, but this doesn't mean that a third party has custom-built that system to perform a set task for its own benefit. Without this being so, the term 'machine' must be defined so loosely as to have no real value. The use of the term machine by Drexler and other mechanists would require that a machine be defined along the lines of 'something that functions in a manner similar to a car', or 'something that functions in a manner similar to a clock', etc. Again, this is to refer to something as a machine by analogy, not through a literal categorical equivalence. A ruthlessly effective tennis player might be described as 'a machine', but no-one has ever presented this as proof of the feasibility of building a machine that plays tennis.

The title of Drexler's most famous book, *Engines of Creation*, makes clear the mechanism of his perspective. Drexler offers a nanoscale mechanical philosophy strikingly similar to the mechanical philosophy of Descartes. In trying to understand the alien landscape of the nanoscale, he employs the same set of metaphors employed by Descartes to understand the alien landscape of the body's interior. Where Descartes found a collection of wheels and counterweights, pumps and levers, Drexler finds a collection of pumps and clamps, cables and fasteners.

This dominant view of nanotechnology sees it as the culmination of a process through which human ingenuity has broken matter down into smaller and smaller components, and reduced its nature to progressively more and more limited principles; once the nanoscale is reached, it is believed that the direct manipulation of a limited number of variables can give total control over matter. Where an emergentist approach argues that complex systems cannot be explained simply through the interaction of isolated component parts, the reductionist approach believes that they can, and therefore that the manipulation of simple, limited components can allow total control over the vastly more complex, system-scale phenomena downstream. The division between chemists and engineers within nanoscale research is simply a new front the mechanists have opened up in the spread of their perspective's dominance.

Perhaps even more significant than traditional mechanism, however, is the suggestion that nanotechnology is also an extension and development of the technologies of information. In the words of Bensaude-Vincent,

> ... Drexler and his supporters have developed a concept of machine that combines an old mechanistic model inherited from Cartesian mechanics – a passive matter moved by external agents – with a more recent computational model of machines inherited from cybernetics. Both the mechanistic model and the cybernetic one rest on the assumption of a blind mechanism operating without intentionality under the control of a program. Biological evolution itself is conceived of as a blind mechanism operated and controlled by an all-powerful algorithm. (Bensaude-Vincent 2006: 16–17)

According to Drexler, 'Molecular manufacturing will do for matter processing what the computer has done for information processing' (1991: 169). While nanomachines are understood to be human-fabricated instances of a larger category of machines that takes in biological life, biological machines have themselves already been cast as materialisations of the informatic workings of nature – naturally occurring information processing units. The key substances linking the two concepts together are RNA and DNA, proteins that supposedly store information and mediate between pure information and its materialisation in life, and that are therefore both emblematic of life as information processing, and matter as composed of molecular machinery. According to Freitas, 'the ribosome [is] the only programmable nanoscale positional assembler currently in existence (Freitas & Merkle 2004: 97–101).

Having given a broad overview of the logic at work within these speculations regarding nanotechnology, I now want to explore these proposed equivalences more fully. However, having introduced nanotechnology largely through an account of Drexler's arguments, there is a danger of caricaturing the field. As illustrated by the Drexler–Smalley debate, Drexler's views are far from the only – or even dominant – views within the nanotechnology community. Indeed, his tendency towards the science-fictional in his speculations has made him a controversial figure, arguably embarrassing to many engaged in nanotechnology research. As noted by Cyrus Mody,

> [A]t least since the founding of the US National Nanotechnology Initiative (NNI) in 2000, Drexler's perspective has continually faced challenges from all of the other stakeholders in the enterprise. Those who seek to make nanotechnology a coherent, well-funded, publicly-supported discipline in the present have tried

hard in the past few years to separate the field from its futurist past. Above all, this means separating it from Drexler, and both prominent and ordinary nanotechnologists have participated in his ritual expulsion in an attempt to mainstream their discipline. (Mody 2006: 109)

However, Drexler's tendency towards overenthusiastic speculation gives his pronouncements value by exaggerating ideas that are present more subtly in other work. While he, like Ray Kurzweil, might be dismissed by some as constituting a 'lunatic fringe' that tarnishes more serious scientific endeavours, the over-exuberant extrapolations such figures make from current scientific understandings throws into relief the logics at work throughout the field. It would be a great mistake to evaluate all nanotechnology research based on Drexler's account, and I am not here seeking to dismiss it wholesale; at the same time, however, I am seeking to suggest that even more cautious and sober understandings of nanotechnology and its potential sometimes share his assumptions to a greater degree than their proponents might like to acknowledge.

Nineteenth-Century Nanotechnology

Despite the futuristic and speculative connotations of nanotechnology and its association with cutting edge or not yet realised technological breakthroughs, fantasies of a nanotechnological future and the means of bringing it about are nonetheless indebted to much older traditions of thought. For example, while the promised wonders of nanotechnology have yet to be delivered in the early twenty-first century, the power of molecular manipulation was already a subject of speculation and fascination in the nineteenth.

The Victorians might have seen evidence to support their faith in the principle of progress all around them – for example in the invention of new technological wonders and the application (and misapplication) of Darwinian evolutionary theory – but not all the new discoveries and theories of the age could be fitted – or forced – into this metanarrative. In particular, while Darwinian evolution suggested an ongoing process of life's spread and adaptation, another great scientific discovery of the age, thermodynamics, presented a limit to the viability of all life. As we saw in the previous chapter, the transfer of energy had come to be understood as the key principle of life, but this life-giving energy was constantly leaking away; the second law of thermodynamics forced Victorian thinkers to confront the idea of an entropic universe, one heading ultimately towards the death of the sun and human extinction (Beer 1989: 159, Keller 1995: 45ff.). However, Gillian Beer has highlighted the potential solace provided by the very agents of entropy; at the molecular scale, the organisation of atoms would remain long after all animal life, and even the celestial bodies that sustained it, were gone (Beer 1989: 172); the stable, elegant structure of the molecule was outside time and immune to entropy. In the words of Victorian physicist and mathematician James Clerk Maxwell,

> [T]hough in the course of ages catastrophes have occurred and may yet occur in the heavens, though ancient systems may be dissolved and new systems evolved out of their ruins, the molecules out of which these systems are built – the foundation stones of the material universe – remain unbroken and unworn.
>
> They continue this day as they were created, perfect in number and measure and weight, and from the ineffaceable characters impressed on them we may learn that those aspirations after accuracy in measurement, truth in statement, and justice in action, which we reckon among our noblest attributes as men, are ours because they are essential constituents of the image of Him Who in the beginning created, not only the heaven and the earth, but the materials of which heaven and earth consist. (Maxwell 1873: 441)

The ability of molecules to survive the extinction of all life might seem like cold comfort (if you'll pardon the pun) to a human being faced with the inevitability of 'heat death', but it also suggested that the key to salvation from entropy lay in the realm of the molecule. To leave the human scale and enter the realm of the molecule would be to 'leave the world of chance and change, and enter a region where everything is certain and immutable' (Maxwell 1873: 440). It was there that the machinery of entropy was in motion and – if only it were possible to act upon that invisible realm – its workings could be turned away from our annihilation. More than a century before Drexler's speculations, Victorian thinkers mused on the possibility of a miraculous entity that possessed wondrous powers by virtue of its ability to interact with matter at the molecular scale.

This fantastical being was originally proposed by Maxwell in the 1860s, and was later dubbed 'Maxwell's Demon' by William Thomson, Lord Kelvin.

> Clerk Maxwell's 'demon' is a creature of imagination having certain perfectly well defined powers of action, purely mechanical in their character, invented to help us to understand the 'Dissipation of Energy' in nature.
>
> He is a being with no preternatural qualities, and differs from real living animals only in extreme smallness and agility. He can at pleasure stop, or strike, or push, or pull any single atom of matter, and so moderate its natural course of motion. Endowed ideally with arms and hands and fingers – two hands and ten fingers suffice – he can do as much for atoms as a pianoforte player can do for the keys of the piano – just a little more, he can push or pull each atom *in any direction.*
>
> He cannot create or annul energy; but just as a living animal does, he can store up limited quantities of energy, and reproduce them at will. By operating selectively on individual atoms he can reverse the natural dissipation of energy, can cause one-half of a closed jar of air, or of a bar of iron, to become glowingly hot and the other ice cold; can direct the energy of the moving

molecules of a basin of water to throw the water up to a height and leave it there proportionately cooled (1 deg. Fahrenheit for 772 ft. of ascent); can 'sort' the molecules in a solution of salt or in a mixture of two gases, so as to reverse the natural process of diffusion, and produce concentration of the solution in one portion of the water, leaving pure water in the remainder of the space occupied; or, in the other case, separate the gases into different parts of the containing vessel. (Thomson 1911: 21–2)

By virtue of its ability to act on individual atoms, Maxwell's Demon could exert power over the most fundamental workings of matter, even nullifying the second law of thermodynamics itself and therefore halting the otherwise implacable progress of entropy that doomed all life to ultimate extinction. Evelyn Fox Keller has already argued that Maxwell's Demon is a key figure in the evolution of a mode of thought that leads to the privileging of DNA as the determinant of life, as it not only attributes importance to the realm of the molecule, but imagines a mechanical and humanlike (rather than supernatural) selective agency at work in that realm (Keller 1995: 55).[9]

As we have already seen, during the 1950s, under the influence of cybernetics and molecular biology, the idea of life as a mechanical process, and particularly as dependent upon the processing of information, was highly influential, and various cyberneticists set about creating artificial 'creatures' whose complex behaviour was determined by flows of information (Johnston 2008: 29–30). During this period, the kinematic self-reproducing automata of John von Neumann, although they were never built, most powerfully foreshadow the mechanical 'life' of the nanobot.

This brilliant Hungarian-born mathematician was obsessed with machine reproduction for much of his career, although the strangeness of the idea meant that it was not always prominent in the work he publicly presented – his friend Norbert Wiener once joked that von Neumann's ideas on machine reproduction might inspire a new Kinsey Report (Freitas & Merkle 2004: 5). Starting at the end of the 1940s and continuing through the 1950s until the end of his life, he toyed with various models of machine reproduction, including the 'kinematic' mode (von Neumann 1966: 82):

The mathematician envisioned a physical machine residing in a 'sea' or stockroom of spare parts ... The machine has a memory tape which instructs it to go through certain mechanical procedures. Using a manipulative appendage and the ability

---

9   Keller argues that Lord Kelvin's musings forge links 'between the animating force of Life and the Demon' (Keller 1995: 57), and that the link Lord Kelvin creates between this imaginary agency and life associated it with vitalism, thus dooming it to unfashionability with vitalism's discrediting. This is ironic, perhaps, given that vitalism's antithesis and final nemesis, molecular biology, also focuses on mechanical workings at an infinitesimally small scale.

to move around in its environment, the device can gather and connect parts. The tape-program first instructs the machine to reach out and pick up a part, then to go through an identification routine to determine whether the part selected is or is not the specific one called for by the instruction tape. If not, the component is thrown back into the 'sea' and another is withdrawn for similar testing, and so on, until the correct one is found. Having finally identified a required part, the device searches in like manner for the next, then joins the two together in accordance with instructions. The machine continues following the instructions to make something, without really understanding what it is doing. When it finishes, it has produced a physical duplicate of itself. Still, the second machine does not yet have any instructions, so the parent machine copies its own memory tape onto the blank tape of its offspring. The last instruction on the parent machine's tape is to activate the tape of its progeny ... (Freitas & Merkle 2004: 7–8)

What is striking about this concept today is its similarity to two things that were either unknown or not fully understood at the time von Neumann conceived it. The first is, of course, cellular reproduction. With the discovery of DNA, the seeming similarity between von Neuman's proposed machinic reproduction and biological reproduction was not lost on those who saw DNA as the 'program' of life (Freitas & Merkle 2004: 6). The other similarity is, of course, to Drexler's nanomachines. In fact, assuming that the kinematic self-reproducing automaton was small enough, it would, simply, *be* one of Drexler's nanomachines. Although nanotechnologists' own origin story casts another 1950s thinker, the physicist Richard Feynman, as the progenitor of the discipline, Drexler's ideas bring together Feynman's concept of building on a tiny scale and von Neumann's idea of machines that collapsed distinctions between machine and organism.

The larger intellectual significance of von Neumann's self-reproducing automata is that they provided a model in which structures of information informed machines and matter, and the combination of information and matter produced machines that shared the attributes of living organisms – so much so that it was hard to draw a distinction between them and living organisms, particularly once it was believed that his models were more-or-less at work in living organisms themselves. More than the idea that machines could be alive, this produced the idea that living organisms were machines, and machines that importantly functioned by processing information.

Self-replication is the quality that is most commonly seen as definitional to life, and therefore the ability to reproduce has been cast as a Rubicon separating 'true' life from artificial approximations. As a result, it is easy to understand why von Neumann was fascinated by the idea of crossing it, even though the practical problems of physical machine reproduction caused him to move away from kinematic automata, largely into the more abstract mathematical realm of cellular automata, and self-reproducing 'artificial life' is today still largely confined to the abstraction of algorithms and computer simulations. Drexler's vision of nanotechnology, however, promises a mechanism by which self-reproducing machines could exist in the physical world – albeit at a very small scale. Furthermore, key to his claims for the viability of this proposal is the belief that the way in which nanomachines could reproduce through mechanisms at the molecular scale is fundamentally the same as that in which organic life reproduces, and is therefore not qualitatively different. However, while the description of self-replicating nanomachines in Drexler's initial account of nanotechnology powerfully captured the popular imagination, more recently it has been dropped from his proposals (see Phoenix & Drexler 2004, Rincon 2004). This is because, while the idea of self-replicating nanomachines did capture the popular imagination, it tended to do so in a way that inspired fear of a future overrun by uncontrollable artificial organisms.[10]

Nanotechnology and Aesthetics

Nanotechnology is an aesthetic pursuit in a fundamentally important way, as human comprehension of and intervention at the molecular scale requires, not just technologies of atomic manipulation, but also technologies that can mediate between the molecular scale and the human senses – primarily human vision.[11] In other words, the machines used in nanotechnology must provide the invisible with an illusion of visibility, creating a 'realistic' representation of something that has no 'real' appearance.

Nanotechnology, as an attempt to manipulate matter at the atomic scale, is the most striking – and perhaps the final – example of the drive to reduce the world to abstract structures of information, and the appearance of new kinds of vision and visual metaphors were key to the birth of the nanotechnology enterprise. Existing as it does outside the realm of direct human perception, it is of necessity understood entirely through human conventions of visualisation and representation and, furthermore, the means of manipulating matter at this scale is provided by the same technological devices that have made it available to the human eye through computer visualisation. Even more than with something like the double helix diagram, then, it's necessary to be aware of just what the images we see of this realm mean. They are visual representations of things that actually have no visual qualities at all, and since relatively little has actually been constructed at the nanoscale, many of the images we see aren't even

---

10 Exemplified by the 2002 Michael Crichton thriller *Prey*, in which out-of-control nanobots set about devouring and assimilating everything around them (Crichton 2008).

11 Vision need not be the only sense addressed by this technology; for example, the nanoManipulator system uses virtual reality to not only create an immersive visual sense of interaction with atoms, but also produces force feedback that creates a sense of physical contours and resistance in the hand of the operator (Guthold et al. 2000: 189).

pictures of things that materially exist; they're just 'artist's impressions' or plans. Nanotechnology as it currently exists, then, is largely an aesthetic endeavour, concerned with finding the right modes of representation to communicate certain ideas to both a specialist and a general audience.

The era of nanotechnology was ushered in by IBM's 1981 invention of the scanning tunnelling microscope (STM) (Schummer 2006: 53–4), a device that allows the precise visualisation of structures at the atomic scale. Again it should be emphasised, however, that the images produced by the scanning tunnelling microscope do not constitute a direct visual revelation of atomic structures; rather, we are witnessing visual information produced by a computer, a simulated perspective produced by the translation of non-visual information into a visual representation.

The scanning tunnelling microscope detects atoms on the surface of a molecule by responding to an electric charge quantum tunnelling between its scanning tip and the material being scanned (Hennig 2006: 145). The tip moves upwards or downwards in response to variations in the voltage caused by different atoms on the surface, thereby producing a kind of 'relief map'. However, this relief map does not result from the direct visualisation of the tip's movement; the presence of some atoms will cause the microscope tip to rise, while the presence of others will cause the tip to drop, meaning that, rather than being pushed upwards by the presence of a particular atom, the topography might actually be pushed downwards. Images of the surface of molecules are therefore often composites of different scans: reversing the polarity of the tip can produce negative contours, which can then be composited with a positive contour so that all atoms are represented by upward deformation, and the different atoms from each scan are then differentiated by 'painting' them in contrasting colours (Hennig 2006). The resulting image of the molecular surface is therefore highly artificial, made by manipulating and seamlessly joining or adding different kinds of information together to create something that satisfies the conventions of human vision. These images often even have been imbued with artificial light and shadow in order to throw their contours into better relief.

Broadly speaking, this aesthetic dimension to scientific research is nothing new. There is a long history of using machines to create new kinds of sensory experience in turn able to produce new kinds of knowledge, and, as we have already seen in Chapter 3, physiology's rise was heavily reliant upon the use of machines, of which Etienne-Jules Marey's chronophotographic rifle (Dagognet et al. 1992: 91–4) is perhaps the most famous. Such machines, memorably described by Lorainne Daston and Peter Galison as 'oracles speaking nature's own language' (Daston & Galison 1992: 116), in almost all cases functioned by taking biological processes that appeared to the human senses as totalities, and fracturing them into isolatable fragments. Thus the beating of a bird's wings or the galloping of one of Leland Stanford's racehorses could be broken up into a collection of components that were connected but distinct through a process of fracturing and fixing in time. Such use of machine vision has been crucial to the development of modern mechanist accounts of nature, as it breaks the unity of the processes of life, rendering them as an assemblage of smaller, interacting components, and mechanism requires an ability to explain material phenomena as composed of smaller, interacting components.

But molecular visualisation equipment like the STM takes this propensity still further, firstly because high-speed photography does not create visual information *per se* – it rather captures visual information unavailable to the human eye; as already noted, the STM creates visual information from a non-visual source. At the same time, however, even this is not, in itself, novel. Other devices used in physiological research, such as the kymograph, or recording drum, allowed the conversion of non-visual information, such as the rhythm of the heart, into visual information amenable to human consideration (Borell 1987: 53).

But there is a second distinction to be made, between the distributive nature of high-speed photography and the aggregative nature of molecular imaging. Rather than breaking a unity into its component parts, molecular imaging takes phenomena characterised by their infinite multiplicity and potentiality and contains them within a unified representation. The atoms of which molecules are composed are in turn composed of parts that are in constant motion and subject to quantum indeterminacy. Not only are nanoscale images visualisations of phenomena bereft of visual properties, but the properties and disposition of these phenomena in space and time defy any interpretation of their co-ordinates informed by human sensory experience. Such visualisation is therefore a highly artificial attempt to render up to human perception a thoroughly inhuman environment where 'there is no light, no color, no gravity, constant motion, blinding speeds and other conditions … A singular entity can be at one location or two at the same instant and never pass through the space in between' (Robinson 2012: 455). As a result, visual representations of nanoscale phenomena are statistical in nature, representing an averaging out of the constant variations in how the electrons at the exterior of an atom appear over time in order to create something visually comprehensible (Tarr & Weiss 2012: 442).

Furthermore, both the scanning tunnelling microscope and the atomic force microscope, which was invented by IBM in 1986, transformed the relationship between observation and manipulation, between the translation and the modification of matter. Moving the probe of such a device close enough to an atom can cause it to 'stick' to the probe, allowing an operator to pick individual atoms up and move them around. This technique allowed the creation of the IBM logo using individual atoms mentioned above, and illustrates another one of the ways in which understandings of the visual get complicated by nanotechnology: the machine that allows us to 'see' this artefact is also the machine that brought it into existence.

The STM and AFM allow the manipulation of atoms, meaning that they are not simply instruments for observing atomic structures, but ones that alter the atomic structures they are visualising. As noted by Peter Galison, 'Imaging, in the nano-domain, has shifted from a form of passive receptivity to an integral part of manipulation' (2006: 173). Galison further argues that nanotechnology problematises the traditional role of science in that, rather than simply seeking to observe and explain features of the physical world, it actually sets itself the task of creating the physical phenomena it seeks to investigate. Again, the subscription to an informatic understanding of matter has led to a breakdown of distinctions between representation and the material object being represented, between metaphor and the object of metaphor, between the translation of physical phenomena into information, and the creation or alteration of physical phenomena through the agency of information. And the invisibility of nanoscale phenomena is key to this breaking down. The STM and AFM both mediate between this invisible realm and the human senses, and how phenomena at the nanoscale are understood is substantially determined by the aesthetic qualities with which they are imbued by this process of rendering the invisible visible.

But there is a further sense in which nanotechnology can be understood as an aesthetic pursuit. The neo-Platonic tradition of Western aesthetics depicts art as mediating between the physical and particular, and the immaterial and universal (Gombrich 1948: 170–71). For Hegel at the beginning of the nineteenth century, aesthetics was the study of how the beauty of art could communicate Truth through the senses; according to Hegel, art was 'the first middle term of reconciliation between pure thought and what is external, sensuous, and transitory, between nature with its finite actuality and the infinite freedom of the reason that comprehends' (Hegel 1993: 9–10). This theme is also apparent in the wider discourse of the nanoscale, in which molecules are frequently understood as the perfect, eternal, and transcendent originators and determinants of messy, flawed, contingent matter, which researchers and their machines translate into visual representations.

Hegel's understanding of beautiful art depended upon a division between the messy, meaningless, material world of the here-and-now and an immaterial world of the absolute, the transcendent, and pure thought and rationality. For Hegel, the beautiful work of art served as a bridge between these two realms, negotiating the gulf between the immaterial and transcendent realm of ideas and the material and particular realm of matter by giving ideas a material particularity. In the words of Keats, 'beauty is truth, truth beauty': we perceive as beautiful that which expresses transcendent truth; therefore beauty is a physical indication of the presence of higher truth. The role of the artist therefore was to create beautiful objects by reproducing reality in a way that stripped away meaningless particularity to expose universal concepts.

It was this relationship between art and higher truth that made beauty an important object of philosophical inquiry at the time: it turned the art museum into a kind of post-Enlightenment temple. In the words of Michel Marra,

> Theoretical developments in the field of aesthetics kept treading over the path of reconciliation between the present world and the world of transcendence through the mediation of the 'beautiful' … Belief in the potential of art to bridge final actuality and the infinite freedom of reason – metaphysical universality and the determinateness of real particularity – made the work of art an epiphany of the Absolute Spirit in front of a newly constituted assembly of believers (the art critics and the appreciators of art) who filled with aesthetics the voice left by theology. (Marra 1999: 5)

In today's terms, this makes art seem like a pretty lofty endeavour, but for Hegel it made it only a lesser form of the struggle to attain insight into the Absolute, a poor cousin to religion and philosophy. The reason for this is that it remains tainted by materiality and particularity. Art is a '"carrying back" into inwardness, in which we do not find the universal thus carried back to its extremest limit to the form of abstract *thought*, that is to say, but is rather suffered to rest half-way at the point in which we find the purely external and the purely inward meet together harmoniously' (Hegel 1975: 213). It seeks to create the perfect marriage of material reality and transcendent concept, but the material dimension of art prevents it from attaining the purity of concept alone. According to Hegel, then, art mediates between the realm of the spiritual and abstract, transcendent Truth and the everyday world of the senses and materiality. It's for this very reason that Hegel believes art to be inferior to philosophy and religion, and something whose time has ultimately passed. This is Hegel's explanation for why, in his opinion at least, civilisation has progressed in every regard except art, which – with the exception of music – has supposedly gone downhill since the Greeks. Beauty arises from the visibility of the Absolute in a physical form that appeals to our senses, but a superior grasp of the Absolute comes from moving beyond sensuous experience completely to a realm of pure thought. Hegel believed that 'art is and remains for us, on the side of its highest possibilities, a thing of the past' (Hegel 1975: 13), something of greater importance to earlier peoples, who had not yet attained a level of civilisation that would allow the kind of rational contemplation he credited to his time.

Philosophy was superior to art because it could leave materiality and particularity behind completely. Art, while inferior to philosophy, was superior to nature because nature did not communicate any higher truth. For Hegel, this meant that any beauty we might perceive in nature – dumb, meaningless nature – could only be of an inferior sort.

Beauty has value in that it directs us away from the specific forms in which we see that beauty. The beautiful art object should capture our attention by seducing us with its form, but we should then move beyond consideration of that form to a contemplation of the higher truth that lies beyond it. ¹

> Beyond a doubt the mode of revelation which a content attains in the realm of thought is the truest reality; but in comparison with the show or semblance of immediate sensuous existence or of historical narrative, the artistic semblance has the advantage that in itself it points beyond itself, and refers us away from itself to something spiritual which it is meant to bring before the mind's eye. Whereas immediate appearance does not give itself out to be deceptive, but rather to be real and true, though all the time its truth is contaminated and infected by the immediate sensuous element. The hard rind of nature and the common world gives the mind more trouble in breaking through to the ideal than do the products of art. (Hegel 1993: 11)

In Hegel's account, artistic beauty seems almost like a necessary evil, whose materiality is required by those without the mental sophistication to deal only with the immaterial and rational. His belief in the march of progress and civilisation led him to believe that such people were growing fewer, and that the role of art had therefore declined.

It might seem like a strange leap from molecules and 3D computer imagery to Hegel's ideas about art, but they are united by a common thematic thread. While Hegel's terminology and explicit objects of concern might be very different, his account of art and beauty is similar in the sense that it depends on a neo-Platonist division between a sprawling, meaningless, devalued material reality and a transcendent, generalised, rational higher truth.

Because nanotechnology doesn't look like anything, its visual significance is completely up for grabs; its representation has a powerful influence over how it is imagined and understood, and it can be represented in any way. In an article dealing with representational strategies for nanotechnology, David Goodsell treats such representations as resulting from decisions by the researcher:

> Two steps are needed to create a picture of a molecule. The first is to find a representational metaphor that captures the salient features of a molecule and presents them in an intuitively obvious way . Choice of these metaphors is a tricky business. Most of the cleverness and artistry of molecular representation is spent, perhaps unknowingly, at this stage … (Goodsell 2006: 45)

The scientist here doesn't seem so far removed from Hegel's artist after all, proceeding through intuition and the clever exercise of artistry. However, Goodsell warns of the 'insidious danger' of images, which can be 'so compelling that they may compete with the scientific facts' (Goodsell 2006: 47).

The public might be misled by uninformed exposure to these representations, but are the 'scientific facts' really opposed to or independent from them? Visual representations themselves arise from the 'artistry' of the researcher, and surely in turn influence researchers' conceptualisation of the phenomena being studied. More than simply an aesthetic consideration influencing the creation of visual representations, the use of metaphors becomes a constituent part of the research endeavour itself.

Given that molecules have no visual qualities, Hegel's beautiful marriage of metaphysical universality and real particularity cannot, strictly speaking, exist at the nanoscale, because visual beauty cannot exist. However, Joachim Schummer has remarked upon the tendency amongst researchers of the nanoscale to attribute molecules with beauty nevertheless (Schummer 2003), and noted the correspondence between the geometric perfection of form in molecular representations and Platonic solids, which performed a similar conceptual role in the Platonic understanding of the Cosmos.

> What Plato explicitly called *the most beautiful bodies in the whole realm of bodies* are not men, animals, plants, landscapes, or whatever kind of natural things the fine arts have tried to imitate. Instead, these are the tiny little bodies, the elemental building blocks of all other bodies in the form of the regular polyhedra. Much too small to be perceived by the senses, they make up the four antique elements: fire/tetrahedra, air/octahedra, water/icosahedra, and earth/ cubes. (Schummer 2003)

Platonic solids prefigured the modern conception of molecules, not only in their posited role as the building blocks of material reality, but also in their representation as perfect, 'beautiful' geometric forms, pure and flawless in contrast to the imperfect and inconsistent shapes of experiential reality.

The immaterial and the material, form and matter: traditional aesthetics is fundamentally concerned with the relationship between these things. In a way, beauty itself is of only marginal concern: it is simply a byproduct of the interaction of truth and matter, rather than a key factor. If the relationship between truth and matter is in order, then the result will necessarily be beautiful; it is therefore not an end in itself.

Such an account of artistic practice and the nature of beauty is likely to be greeted with a great deal of scepticism today. Is there such a thing as universal, transcendent truth? Is there a universal conception of beauty? Does all art need to be beautiful? Many people would say no to the latter question, and most would say no to the former two. In Chapter 1, Ramachandran and Hirstein, while actually saying yes to the last two questions, would of course explain the existence of beauty in art as resulting from innate neurological factors, rather than the calling into material form of transcendent truth.

In-Forming Matter

So where does this leave aesthetics? With no transcendent, immaterial truth, this understanding of beauty is unworkable. But do we really live in a world largely free of belief in transcendent truth? After all, science searches for an elegant set of principles underlying the particularity of matter. Aristotle understood form to be closely tied to function – the purpose of an object and the way in which it functioned in the world were part of its transcendent nature, over and above the question of whether it was ever called upon in the particularity of an individual, material form (Aristotle 1986: 15). The project of understanding the fundamental nature of the physical universe similarly seeks to identify the underlying principles existing above the contingencies of actual matter. Pythagoras believed that material phenomena could ultimately be reduced to mathematical principles, and in laboratories around the world computers are working through numerical data representing natural phenomena in the hope of discovering some transcendent, reductive principles behind nature. The fMRI machines that provide data about brain function to researchers like Ramachandran reduce the brain to cascades of digital information, which are then reconstituted as a 3D, visual representation of living human anatomy. Digitalisation converts analogue particularity into immaterial, transcendent, numerical immateriality; research into genetics looks for the hidden code that gives rise to the material specificity of the living body. In other words, the reductive method and various other ways of generating, processing, storing, or understanding material phenomena – in short, the whole process of informatic dematerialisation discussed in the previous chapter – parallels the work of Hegel's art, which, 'by comparing what is otherwise stained and rent through the contingent elements of external existence with the harmony that is essential to its notional truth, rejects that in the world of appearance which it is unable to combine in such a unity' (Hegel 1975: 212), recreating the world in a purer, truer form than the dishevelled agglomeration of things produced by nature. It was the Greeks who first put forward a theory of atoms as the invisibly small building blocks of reality that serve as points of articulation between Pythagoras's mathematical universe and physical reality, entities that 'were so small, [that] they could not be experienced through the senses. The only way to gain any knowledge of them was through the rational powers of the mind' (Channell 1991: 14–15). These ideas, rediscovered and popularised during the Renaissance, inform the reverential posture later taken towards atoms by the likes of Maxwell.

The process of dematerialisation pulverises material specificity and reconstitutes it as a generalised, departicularised abstraction. The reductive, dissective method – from the destruction of the unique, idiosyncratic architecture of particular human bodies to produce standardised, universally applicable anatomical representations onwards – is about the loss of material

174

specificity and uniqueness. Hegel would, presumably, approve of all this, seeing it as further confirmation that the sophisticated modern mind does not need to wrap its truth in a seductive physical form, but can grapple with it on its home turf, in an abstract cognitive plane. Central to the historical accounts of Jonathan Sawday and Barbara Maria Stafford is the idea that varied pursuits and modes of thought were implicated in one another in just such a way; that anatomy and biology and fine art and poetry from the Renaissance to the nineteenth century – at least – shared a set of values and world view, one which believed the dissection, stripping down, atomisation and abstraction of nature to be the route to the discovery of higher, generalised truths.

Ironically, the progression towards generalisation and abstract rationalism seen in Hegel's aesthetics has continued even in the absence of a belief in the absolute Truth he saw as its object. Whether in art or science, today it is not commonly believed that we are labouring to attain an understanding of some transcendent, ultimate truth of the kind envisioned by Hegel. So what is the nature of this information that is believed to hide behind physical reality, waiting for us to sift it from the dross of contingency and ugliness? A conventional scientific answer to this question would most likely be that it is simply understanding that is being sought. Translating the 'code' of DNA won't tell us about God because he is no longer credited with drawing up the blueprint for the human body, but it will provide us with new information about the nature of the human body. There are various possible benefits in this, of course, but the one that is most prized, and has captured the popular imagination most of all, is the possibility of actually changing the nature of our physical selves through genetic engineering. For Hegel, the marriage of truth and matter in the beautiful artwork should function to seduce us with its sensuous form only to then direct us away from its form to an immaterial idea, 'away from itself to something spiritual which it is meant to bring before the mind's eye' (Hegel 1993: 11). While the later attempt to reduce reality to information similarly understands the material specificity of the world as simply a product of deeper structures of information, this information is no longer *about* some higher, ultimate truth, as no such higher, ultimate truth is believed to exist. The information is now an end in itself. The quest for information no longer understands information to be part of some meaningful message; the circulation of information as an abstract, immaterial quality is simply understood to be a fundamental, underlying property of the universe we live in. If this information is to be credited with any meaning or significance, this can only be done by applying it back to the material specificity from which it was originally extracted. The human body can be reduced to a collection of genetic data, but this genetic code has significance only if it is used to refer back to the complexity of the human body it was used to simplify. As a result, while Hegel's artistic beauty was part of a linear trajectory that took the human mind from the meaningless shambles of material reality to a realm of the Divine, the Absolute, and pure reason, the secular quest for truth follows a circular trajectory that similarly seeks to escape material specificity by abstracting out pure, immaterial information but, having done so, can only then fall back into the realm of the material because it lacks any higher realm to which this can be applied.

Having broken matter down to a molecular scale, there is really only one direction left to go: back up. Having disassembled matter to the level of molecules and then atoms, the goal now is to start putting these components back together, recreating the material world in a form more pleasing, more meaningful, and more rational for human beings. The nanotechnologist has replaced Hegel's artist, seeking to reproduce material reality in a way that harmoniously combines matter and information. Instead of searching for an absent God through the discovery of hidden information, human beings hope to replace the missing deity by transforming themselves into Demiurgic masters of matter. Fantasising about remaking the human body through the manipulation of genetic code or even the remaking of all matter through the shuffling of molecules, the abstraction and devaluing of materiality leads to a renewed concern with materiality. Once the informatic view of the world has attained complete dominance, matter is no longer seen as antithetical to information, and therefore is no longer something to be rejected. When everything – the human body, trees, rocks … everything in existence – is understood to be nothing more than a collection of invisible informatic structures, there is no longer any distinction to be drawn between the material and the immaterial. The whole world is information, and our interactions with it can be reduced to the 'programming' of matter.

In his book *Biomedia*, Eugene Thacker has provided a detailed treatment of various new biomedical technologies, and the ways in which they extend this traditional understanding of the relationship between information and matter in paradoxical new directions:

> This confluence of information and matter is an interesting implication, for it works against a long tradition of technical thought that dissociates information from its material substrate. From a certain perspective, nanotech can be seen to be embodying information in molecules; or rather, positing the indissociability of matter from data … (Thacker 2004: 138)

Nanotechnology's promise of miraculous 'smart materials' (Küchler 2008: 105) suggests that, since all matter is made up of information systems anyway, we just need to engineer matter to function as information technology useful to human beings. We can build clothing that functions as a computer, or build our houses with walls that function as video screens – at the most basic level there is no difference between matter and information, biological or organic substances and machines, so we can use matter to process our information, and use organic substances as machines.

Where N. Katherine Hayles's examples of virtuality demonstrate the propensity to see material reality as penetrated by flows of information (see previous chapter), the nanotechnology example does not create the same sense that information is opposed to materiality. Rather than matter being devalued by a focus on information, matter comes to be seen as *itself* information. Once it is accepted that matter is only an arrangement of information anyway, the two terms can be imagined as harmoniously integrated, with human beings intervening only to make the relationship more fruitful. Matter is already an agglomeration of information, but perhaps new technological advances will allow us to stuff even more information into matter, or choose the kinds of information for which matter serves as a receptacle. Ultimately, everything becomes a system processing information.

The extremes of this thinking are apparent in some of the more speculative work on artificial intelligence, such as that of Hans Moravec and Ray Kurzweil (see Black 2013). Moravec's chart that supposedly plots the relationship between different information systems such as DNA and the US Library of Congress, calculators and bees, personal computers and sperm whales, the telephone system and the human brain (see Introduction), depends upon the idea that information processing is a fundamental attribute of matter. Ray Kurzweil even asks:

How Smart Is a Rock?

To appreciate the feasibility of computing with no energy and no heat, consider the computation that takes place in an ordinary rock. Although it may appear that nothing much is going on inside a rock, the approximately 10^{25} (ten trillion trillion) atoms in a kilogram of matter are actually extremely active. Despite the apparent solidity of the object, the atoms are all in motion, sharing electrons back and forth, changing particle spins, and generating rapidly moving electromagnetic fields. All of this activity represents computation, even if not very meaningfully organized. (Kurzweil 2005: 131)

This view of computational matter arises from MIT professor Seth Lloyd's theory that the Universe is one vast quantum computer, put forward in his book *Programming the Universe* (2006), although he gives credit for the idea 'that all physical systems register and process information' (Lloyd 2002) to none other than James Clerk Maxwell, father of the famous Demon, arguing that Maxwell and others originated information theory in their efforts to explain the activity of molecules, their ideas then going on to influence cyberneticists such as Norbert Wiener (Lloyd 2006: 69–70, see Wiener 1961: 37).

The universe is the biggest thing there is and the bit is the smallest possible chunk of information. The universe is made of bits. Every molecule, atom, and elementary particle registers bits of information. Every interaction between those pieces of the universe processes that information by altering those bits.

That is, the universe computes, and because the universe is governed by the laws of quantum mechanics, it computes in an intrinsically quantum-mechanical fashion; its bits are quantum bits. The history of the universe is, in effect, a huge and ongoing quantum computation. The universe is a quantum computer.

This begs the question: What does the universe compute? It computes itself. The universe computes its own behavior. As soon as the universe began, it began computing. At first, the patterns it produced were simple, comprising elementary particles and establishing the fundamental laws of physics. In time, as it processed more and more information, the universe spun out ever more intricate and complex patterns, including galaxies, stars, and planets. Life, language, human beings, society, culture – all owe their existence to the intrinsic ability of matter and energy to process information. The computational capability of the universe explains one of the great mysteries of nature: how complex systems such as living creatures can arise from fundamentally simple physical laws. (Lloyd 2006: 3)

This is the most extreme example of the familiar abstracting manoeuvre that substitutes representation for reality: the human-made, abstract representation is claimed to come first, with the material, physical phenomena it represents arising as a product of the originating representation. Information, as an abstract way of describing material phenomena, is credited with an objective reality independent of material phenomena and so, because all material phenomena can be described in terms of information, information is reasoned to be a universal animating principle behind all material phenomena. Lloyd's argument for the Universe as quantum computer very clearly follows this circular process whereby description is claimed to produce the thing described: to perfectly simulate the entire universe would require a quantum computer the same size as the Universe, which would mean that the quantum computer and the Universe are the same, meaning that the Universe is the same as a quantum computer, in turn meaning that the Universe *is* a quantum computer.

A quantum computer that simulated the universe would have exactly as many qubits as there are in the universe, and the logic operations on those qubits would exactly simulate the dynamics of the universe … Such a quantum computation would constitute a complete description of nature, and so would be indistinguishable from nature. Thus, at bottom, *the universe can be thought of as performing a quantum computation.* Likewise, because the behavior of elementary particles can be mapped directly onto the behavior of qubits interacting via logic operations, *a simulation of the universe on a quantum computer is indistinguishable from the universe itself …*

[T]he universe is nothing but bits – or rather, nothing but qubits. Mindful that if it walks like a duck and quacks like a duck then it's a duck, from this point on we'll adopt the position that since the universe registers and processes information like a quantum computer and is observationally indistinguishable from a quantum computer, then it *is* a quantum computer. (Lloyd 2006: 154)

This entire argument is a case of begging the question: the 'quantum computer' Lloyd argues could simulate the Universe is simply an identical copy of the Universe – it can only be claimed to be a quantum computer if the Universe itself has already been accepted to be a quantum computer.

The informatic view ultimately treats all material phenomena as manifestations of the transcendent stuff of information. Therefore, at base, bees, computers, books, people, and even rocks, are fundamentally the same; they are just different particular manifestations of the same primary substance. According to Lloyd, the entire Universe is simply a gigantic quantum computer, endlessly processing information. Since information theory has jettisoned meaning, there is no need for such supposed processing to have any purpose or product; any arrangement or organisation of anything can be translated into information. Any changes in the state of matter can count as information processing and, since – at an atomic level – all matter is endlessly changing state, all matter is information processing.

At this point, the information processing model explains far more than just the genetic blueprint of the human body or the immateriality of the human mind – it explains literally the entirety of reality. The computer becomes a machine model for not just brain, but rocks, leaves, and indeed the entire Universe as a holistic entity. It does this by articulating information and matter more closely than ever before, and suggesting that – when examined at a fine enough grain – even matter is information. At the molecular scale, even the supposed fixity, stability and particularity of matter itself dissolves – everything is just arrangements of a fixed number of generic, universal components that are themselves dynamic and unstable. When information becomes a science, it presents a scientific account of the mechanism by which form determines matter.

Dissecting Reality

The etymology of the word 'information' associates it with the giving of form, but today information is broadly considered to *be* form, in that it occupies the form position in a division between matter and form. Matter is the expression of form, the raw, inert substance shaped by form, which is now understood in informatic terms. Information is abstract, immaterial and transcendent, manifesting itself in the particular through the patterns seen in material structures. Behind the complexity of the human body lies a pattern of information existing outside the human scale, expressing itself in the particular through material patterns at the molecular scale; my DNA is one iteration of an evolving information pattern, the computer program that runs human life.

The elegant structure of molecules themselves (at least that which is conferred by the neat diagrammatic visual representations of human beings) is itself an expression of an atomic code, the computer program that runs material reality.

Nanotechnology represents an important historical development in the inexorable rise of information. It is a continuation and expansion of information's colonisation of a wide variety of popular and scientific understandings of the world, but perhaps represents a point from which further gains are not possible, because it destroys the very dichotomy on which it depends. In the words of Eugene Thacker,

> Nanotechnology is a unique type of posthumanism because, unlike AI (artificial intelligence) or AL (a-life), its primary concern is not so much 'mind' or abstract 'pattern' but rather the body viewed at the atomic-molecular level. While adopting an informatic worldview consonant with much posthumanist thinking, nanotech research leads not to increased informatic abstraction (uploading software minds), but rather, in a kind of rematerializing loop, back to the body (the body of the very small). (Thacker 2004: 137)

In previous moments in the development of these ideas, one term – form, structure, Truth, spirit, information – was always privileged and often cast as the first cause to the other – matter, particularity, etc. In the age of information, the power of information as a concept depends precisely on the idea that the material and particular can be reduced to information, allowing all the messy, disorganised contingencies of the world to be captured in diagrams and formulae. Nonetheless, of course, like the distinction between form and matter, it depends upon an opposition between the two terms, and the idea that, at a fundamental level, one cannot be reduced to the other.

Nanotechnology, on the other hand, puts forward the idea of a single 'programmable matter' (Thacker 2004: 123), wherein one term *can* be reduced to the other. Acting on matter at an atomic scale produces the idea that matter itself is a code: atoms are a fixed, finite and predictable set of components that give rise to all matter simply through variations in their arrangement. That is, my heart, a leaf, a spiderweb and a diamond are all composed of the same carbon atoms; the only difference is their 'syntax' – the way in which these carbon atoms are arranged and combined. Molecular manufacturing promises ultimate control over matter through the ability to edit that syntax, modifying the code underlying matter in order to change it in whichever way we desire. This idea reduces matter itself to a kind of information system. Where the DNA molecule is seen to determine the nature of life, molecules more generally become the 'DNA of matter', and just as the informatic view of DNA leads to the idea of altering life by reprogramming its underlying genetic program, the informatic view of all molecules leads to the idea of reprogramming matter of all kinds – both animate and inanimate. In fact, following the informatic mode of thought to its conclusion, there is no distinction to be made between the two. Cells become nothing but naturally occurring nanobots, invoked as proof of concept of man-made nanobots. Nature is transformed from a blind watchmaker to a blind nanotechnologist – appropriate given that vision does not exist in the nano-realm anyway.

Like Hegel's art, nanotechnology articulates the abstract and transcendent with the contingent and material, producing physical manifestations of the ideal. Hegel believed that art combined 'metaphysical universality with the determinateness of real particularity' (Hegel 1993: 25), while Ray Kurzweil believes nanotechnology to be '[t]he intersection of information and the physical world' (Kurzweil 2005: 226). The key difference between how these two sets of terms are understood to relate is that Hegel's art should ultimately carry the human mind away from the material into the immaterial realm of Truth, while nanotechnology represents a kind of looping back of the abstract into the material. While the informatisation of matter initially followed the same trajectory away from the material, with no transcendental realm available to it, it can only return to the devalued realm of materiality. This leads to the characterisation of matter itself as information, obliterating the opposition from which the trajectory was launched and creating a sense that there are only different forms of information. The material regains a sense of importance and value, but only as a result of being seen as nothing more than information itself.

At the molecular scale, an informatic account can be given for how information determines matter. Since the late 1940s a much older system of thought dividing form and matter, the transcendent and the immanent, from one another has been scientificised and energised through technology, allowing one side of the division – information – to claim more and more territory, and leaving the other – matter – seeming progressively more and more worthless and redundant. A key early victory was the ability of genetics and computers to cast the body – at an earlier intellectual moment understood to be a stubborn and inescapable bastion of the base and material – as itself most importantly an information system through cybernetics, molecular biology and material accounts of the mind. The influence of information processing models has since continued to spread until, in the discourse of nanotechnology, belief in the common origin of all matter in information processing ultimately renders any meaningful distinction between living body and machine redundant; both are simply physical expressions of structures of information.

Conclusion

The human body is probably the feature of our world we find most familiar and readily understandable, and simultaneously most strange and unknowable. Throughout history, the body has been utilised as a template for understanding other things (often by seeing it as a microcosm of some larger entity such as society or the heavens), but, given the body's own mysteries, this has also introduced a reciprocity between our understanding of body and world. This reciprocity creates a paradoxical kind of metaphor, in which no term is privileged in relation to the other. Rather than there being a first term whose qualities are attributed to the second, there are two second terms that are used to inform understandings of one another without either functioning as the originator of those qualities. This produces a looping effect, where each term refers back to the other, and the attribution of particular qualities to one term is legitimised by the attribution of the same qualities to the other term, but neither has the authority of being anchored to something outside the loop.

The reciprocal relationship between body and machine is a particularly influential and long-lived example of this. In Chapter 1, I suggested that the tendency to project the attributes of bodies out into our environment results from innate features of human perception, and the history of the body–machine relationship certainly serves as an example of this. However, while the fixation on machines in relation to the body reflects a particular significance for this particular kind of object, that should not be taken to mean that there is an innate human tendency to focus on machines specifically.

There are particular reasons why machines are well suited to the attribution of corporeal qualities, and bodies are well suited to the attribution of mechanistic properties: both are notably characterised by movement, and machines tend to be designed specifically to reproduce the actions of bodies. At the same time, however, the reciprocal understanding of bodies and machines has evolved far beyond a reliance on such obvious similarities, and the shared qualities seen as key to the relationship have changed over time. The mechanism of Descartes, for example, is a far cry from that of modern cognitive science.

And yet, while both Descartes (were he alive today) and cognitive scientists might highlight the points of difference between them, there are also key similarities. The looped metaphor of body and machine has not sprung up anew over and over again with successive technologies – like technology itself, it is always appropriating and redeploying what has come before. Modernity has nourished this relationship with a steady stream of new machines – as the attribution of lifelike qualities to one machine fades with demystifying familiarity, a new, unfamiliar machine with seemingly greater animation and autonomy is never far away. But it is also the case that an existing commitment to a framework of understanding based on the body–machine relationship has encouraged its perpetual re-justification even as understandings of what bodies and machines are have changed. Again, key to the success of this relationship is precisely the fact that neither body nor machine has served as a stable category with which the other term must comply. As the attributes of machines have changed, understandings of bodies have also changed to match them, maintaining the integrity of the relationship even as qualities previously seen as necessary to it have disappeared over time.

Key to the initial forging of a connection between machine and body is an understanding that the body is composed of various subsystems that perform various tasks, such as pumping, transferring energy, or moving limbs. The practice of dissection highlighted this aspect of the body by disassembling bodies into these systems and examining them in isolation. The body thus investigated was notably bereft of animation, but the cultural setting within which Renaissance anatomisation took place nonetheless provided the means to fill this gap. To describe the body as a machine in sixteenth and seventeenth century Europe was logically consistent with a belief that the human body had a designer and creator and had been assembled for a purpose. Just as human beings designed and built clever machines in order to act upon the world in particular ways, God had designed and built the human body for his own purposes, and it was God who wound the mechanism and set it in motion. This framework of understanding is, of course, most clearly and influentially set out by Descartes, who states that the material body is a divinely built contrivance available for the use of an immaterial soul, which, while dependent upon the body to act upon the world, is nonetheless of another order entirely and can exist independently of it (for example after death).

However, the seeming impossibility of providing a convincing account of how the material and immaterial are joined gave rise to entirely materialist accounts, in which the bodily mechanism is claimed to be the only component of the self. This position, however, gives rise to a different problem: that of providing a mechanistic account of Descartes' stuff of thought. La Mettrie claims that psychological states can be accounted for simply in terms of physiological changes in the body (La Mettrie 1927: 11–12), but this is hardly convincing when dependent on the physiological knowledge available in his day. The physical movements of the lungs or heart, or the forces exerted by the arm, might have invited analogies with the operation of machines, but there was clearly no existing machine that generated thought, language or memory. It was this very gap that had allowed Descartes to draw his conclusions regarding the irreducibility of thought to material causes, which he presented as the guarantor of his philosophical method.

Automata makers such as Jacques de Vaucanson sought to amaze the public with demonstrations of just how lifelike machines could seem, but he did not claim to have created life. Automata demonstrated an ability to perform tasks previously seen as solely the domain of the ensouled human being, such as writing or playing music, but these mechanical creations can be understood as an effort to map the border separating living body from machine by testing the limits of what machines could and could not do, rather than erase that border and establish the interchangeability of the two. Nonetheless, there is a clear reciprocity between body and machine here: a machine could be a kind of body, and building such a machine, it was believed, could increase understanding of the living body.

By the nineteenth century, however, enthusiasm for this approach had faded. Both autopsy and automaton were considered unsatisfactory paths to unlocking the secrets of the body because they lacked the dynamism and energy crucial to life. However, the mechanisation of physiological studies promised to capture the dynamism of living processes where anatomisation could not. In so doing, it only rarely produced machines that appeared to *be* living bodies, as automata-building had done, but it did bring a new intimacy to the relationship between body and machine. Machines followed the rhythms and flows of bodily processes, recording secrets hidden even from the body's owner. Not only did this lead to the creation of machines calibrated to and organised around the body, but it also rendered the body more profoundly in machinic terms. The objective truth of what the body was and how it functioned was believed to be available only to the inhuman perception of machines, who were untainted by human perception's investment in bodily forms. These investigations created an exchange between a machine-like body and the machines that recorded it, an exchange taking place outside human perception or awareness; a human observer could only inspect the traces left by the machine after the exchange was complete. Successive technologies of visual imaging, from the microscope and the X-ray exposure to the high-speed camera, made available to human vision views of the body that were alien to human perception, shifting temporal or spatial scales, or even up or down the radiation spectrum, beyond the reach of the human eye. These new ways of seeing the body, being produced by machines, reduced the living body to a set of machinic processes available only through the machinic processes of the imaging technology itself.

At the same time, a new intimacy was coming to the relationship between body and machine elsewhere. While clockwork automata had been demoted from philosophical experiments to toys, newer machines were creating new ways of thinking about bodies in mechanistic ways. On the factory floor, industrial machines were replacing the movements of human bodies and human bodies were being ordered and rationalised on the model of the industrial machine, and the bodies of individual workers were increasingly integrated into larger assemblages of machine and body. Furthermore, where the clockwork mechanism had stored energy provided by a human operator and then expressed it until it was exhausted, the steam engine and later motors produced their own energy from fuel in a way analogous to the living body. The mechanical measurement of bodily processes and the integration of bodies into mechanical assemblages met in the appearance of a 'science of work' intended to make the most efficient use of the machine that was the worker's body and its productive energies.

The idea of the worker as machine and machine as worker produced the figure of the robot, in some ways a successor to the clockwork automaton but also importantly different. Rather than a philosophical experiment aimed at recreating attributes of the living body so as to further understanding, the robot begins with a belief in the possibility of mechanically recreating attributes of the living body and seeks to use this to free the human worker from labour. However, the robot is less effective as an illustration of the shared attributes of living body and machine than as an illustration of the abiding faith in and fascination with the possibility that this might be true. While real robots do exist and work in robotics has gained ground, the robot has been far more successful as a mythological, rather than practical, figure. Even today, claims of the success and future promise of robotics continue to far exceed the available evidence.

In its promises regarding a revolutionary future, robotics is allied to artificial intelligence, and shares its desire to move beyond the creation of machines that externally reproduce human bodily capacities to machines that can actually attain a kind of thinking, conscious, personhood. Robots can be – and are – produced to serve useful functions without attaining such a level of sophistication, but the promises for the future made by the discipline and its hold on the popular imagination primarily circulate around this idea. Indeed, the lack of clarity concerning the relationship between machines that simply imitate human behaviour and machines that can think is responsible for a tendency to misunderstand developments in the field. Honda's ASIMO robot is an excellent example of this: popular accounts of this robot treat it as though it has some kind of self-sufficient agency, but in reality it has no independence at all and is simply directed by a human operator via remote control.

The boundary is clearer with artificial intelligence, given that it seeks to reproduce Descartes' posited immaterial, non-corporeal component of the human self, and yet this boundary is still not terribly clear for all that. First, the term artificial intelligence is now applied to a wide variety of software tools that are clearly not conscious and do not approach human intelligence despite having been generated by artificial intelligence research. Second, the very fact that the AI movement began by setting itself the task of creating something immaterial, non-corporeal, and in reality beyond any widely shared, practical definition, could not help but make the boundaries of its concerns unclear.

The AI movement is an undertaking dependent upon the powerful shift in the body–machine relationship brought about by the appearance of the computer. The computer has provided a solution to the problem of a mechanistic account of thought, the problem that left Cartesian mechanism with the irreconcilable excess of the *res cogitans* and rendered previous attempts to do away with it, like that of La Mettrie, unsatisfactory. Computationalist accounts of human cognition claim to have finally done away with Cartesian dualism, being able to provide a purely mechanistic and material account of all aspects of human bodily life and experience. Where previous machines could provide models for how the rest of the body worked, the computer, it is claimed, now provides a model for how the brain works to produce a mind.

However, an unacknowledged dualism remains in such accounts, and the computationalist model retains its own 'ghost in the machine', a ghost made from information. 'Information', understood as an immaterial, transcendent substance, is substituted for Descartes' *res cogitans* in the computationalist account. The mind remains as something immaterial and, in fact, computationalist accounts tend to drive mind and consciousness into the realm of immateriality by cutting them off from the body and the environment. Computationalist approaches to neuroscience employ many of the machines and technologies available for mechanistically and objectively investigating the body discussed in this book – dissection, fMRI scans, etc. – in an attempt to objectively explain the mind and consciousness, but the mind and consciousness, precisely by virtue of being subjective phenomena, are invisible to these technologies, just as our holistic, invested, subjective perception of other living bodies is invisible to them.

Faith in 'information processing' results from the development of digital computing, and the division between hardware and software. Materiality is devalued as simply the contingent physical instantiation of a system necessary to support the important business of information processing. The mind comes to be seen as a program running on the software of the brain, an immaterial phenomenon hovering above the base flesh. The rest of the body and the environment in which it is situated are believed to be separated by an unbridgeable gulf, phantasmatically experienced by a self imprisoned in the Plato's Cave of the skull. If the brain were to be disconnected from the rest of the body, it is claimed, it could continue its information processing in isolation as long as it were supplied with some alternative means of input; and even the brain itself could be discarded as long as some equivalent information processing system were available to continue running the computer program of the mind. The base flesh is not only a producer of information but also its result, being understood as produced by information-processing going on at the molecular level in DNA.

Molecules in general have since Classical Greece been seen as the hidden, unchanging, pure and perfect underlying building blocks of reality. Seth Lloyd credits the creation of information theory to the Victorian physicist and mathematician James Clerk Maxwell's work on molecules (Lloyd 2002: 69–70), and the idea of molecules as the invisible informatic agents behind history and material reality has developed first through the privileging of the DNA molecule as the designer and creator of life, and then through utopian speculation regarding the destiny of nanotechnology. Where a privileging of information processing led to a focus on the immaterial, including the mind and 'intelligence' as phenomena constructed from information and therefore separate from the body, the prospect of manipulating atoms and acting at the molecular level, introduced with the invention of the scanning tunnelling and atomic force microscopes in the 1980s, led to a belief that even matter resolved itself into information structures when handled at a small enough scale. Molecules are cast as the 'genes' of material reality, and changing their syntax will allow matter itself to be reprogrammed. What appears to be inanimate matter at a coarse grain turns out to be, at a fine grain, patterns of information – the entire Universe is a quantum computer that has been computing reality since the Big Bang. The division between information and materiality, form and matter, overbalances in favour of information until the entire dichotomy collapses on itself. With information having colonised everything in the Universe, there is no longer any other term to oppose to it.

The living body was the first territory captured in this conquest through the application of information theory to molecular biology. The human flesh, traditionally seen as the intractable burden borne by an immaterial soul or the necessary machine used by the mind to act upon the world, came to be considered a construction of immaterial information at the molecular level. This did not reconcile matter with the immaterial – on the contrary, it made the material seem even more unworthy of consideration, given that it was simply the product of originary information processing, and the entire process could be understood and even controlled at the information processing level. Similarly, AI initially believed in the possibility of creating a mind as an immaterial software construction independent of any bodily or environmental interactions. Now nanotechnology promises the reprogramming all matter as genetic engineering once promised to reprogram the nature of human beings, returning the informatic view's attention to the realm of the material. However, once again this renders materiality simply a byproduct of privileged information structures.

Machines are artefacts that, by definition, must effectively perform some function in the material world. As a result, while these ways of thinking derived from body–machine interactions might be historically contingent, machines are nevertheless required to reliably and objectively demonstrate certain desired material properties. Understanding a machine with reference to a particular conception of the body cannot objectively alter how the machine functions; similarly, how a particular kind of medical care or physical training affects the body will not be determined by the body's posited relationship with machines, but largely by the body's material properties. Nonetheless, what kinds of projects or investigations arise regarding bodies or machines will be guided by these ideas, and this is particularly apparent in attempts to create new kinds of technologies or design new kinds of machines. For example, at an earlier moment the power of genes over the development and behaviour of the body was greatly exaggerated (see Keller 2000: 144–8), and this was clearly the result of extrapolating from broader informatic frameworks of understanding in order to sketch in areas that were still unclear or identify future research possibilities. Speculation regarding the future development of nanotechnology is in a similar position today, using much older mechanistic ideas about how matter and information function to imagine its future direction. Perhaps these extrapolations will turn research in productive directions, or perhaps not, but either way they will have an influence over the field of potential inquiries available. Robotics and Artificial Intelligence remain the greatest illustrations of the power of such frameworks of understanding: perhaps one day there will be thinking, human-like machines in the world, or perhaps this dream will never be realised, but the dogged refusal to abandon this dream despite so many failed strategies and predictions demonstrates the degree to which it is driven by non-scientific fantasies, desires and assumptions, rather than led by scientific breakthroughs. When we consider our long history of projecting the human body out into the features of our environment, this desire to literally transform the objects around us into bodies is hardly surprising.

Bibliography

Anon. 2005. *BBC – Science and Nature – How Art Made the World*. [Online]. Available at: http://www.bbc.co.uk/sn/tvradio/programmes/howart/ [accessed: April 26, 2013].

Anon. 2005. *Q3 FY2005 Sony Group Earnings Announcement*. [Online]. Available at: http://www.sony.net/SonyInfo/IR/info/presen/05q3/qfhh7c000008adfe.html [accessed: July 2, 2013].

Anon. 2006. *Body Worlds*. [Online]. Available at: http://www.bodyworlds.com/en.html [accessed: June 21, 2013].

Anon. 2008a. *BBC Two – Horizon, 2007–2008, Where's My Robot?* [Online]. Available at: http://www.bbc.co.uk/programmes/b00g4ypg [accessed: April 26, 2013a].

Anon. 2011. *Singularity University Website*. [Online]. Available at: http://singularityu.org/ [accessed: July 2, 2013].

Anon. 2013a. *ASIMO Website*. [Online]. Available at: http://world.honda.com/ASIMO/ [accessed: July 2, 2013].

Anon. 2013b. *NNI Budget*. [Online]. Available at: http://www.nano.gov/about-nni/what/funding [accessed: June 5, 2013].

Aristotle. 1986. *De Anima*. Harmondsworth: Penguin.

Asendorf, C. 1993. *Batteries of Life: On the History of Things and their Perception in Modernity*. Berkeley: University of California Press.

Avatar (dir. James Cameron, 2009).

Babbage, C. 1994. *Passages from the Life of a Philosopher*. London: William Pickering.

Baird, D. Nordmann, A. and Schummer, J. 2004. *Discovering the Nanoscale*. Amsterdam: IOS Press.

Bakhtin, M. 1984. *Rabelais and His World*. Bloomington: Indiana University Press.

Baudrillard, J. 1996. *The System of Objects*. London and New York: Verso.

Bedini, S.A. 1964. The role of automata in the history of technology. *Technology and Culture*, 5(1), 24–42.

Beer, G. 1989. 'The Death of the Sun': Victorian solar physics and solar myth, in *The Sun is God: Painting, Literature, and Mythology in the Nineteenth Century*, edited by J.B. Bullen. New York: Oxford University Press, 159–80.

Bensaude-Vincent, B. 2006. Two cultures of nanotechnology?, in *Nanotechnology Challenges: Implications for Philosophy, Ethics, and Society*, edited by J. Schummer and D. Baird. River Edge, NJ: World Scientific, 7–28.

Bensaude-Vincent, B. 2007. Nanobots and nanotubes: Two alternative biomimetic paradigms, in *Genesis Redux: Essays in the History and Philosophy of Artificial Life*, edited by J. Riskin. Chicago: University of Chicago Press, 221–36.

Black, D. 2011. What is a face? *Body and Society*, 1(4), 1–26.

Black, D. 2013. The digital soul, in *Handbook of Research on Technoself: Identity in a Technological Society*, edited by R. Luppicini. Hershey, PA: IGI Global, 157–74.

Black, D. 2014. An aesthetics of the invisible: Nanotechnology and informatic matter. *Theory, Culture & Society*, 31(1), 99–121.

Bohde, D. 2003. Skin and the search for the interior: The representation of flaying in the art and anatomy of the Cinquecento, in *Bodily Extremities: Preoccupations with the Human Body in Early Modern European Culture*, edited by F. Egmond and R. Zwijnenberg. Burlington, VT: Ashgate, 10–47.

Borell, M. 1987. Instrumentation and the rise of modern physiology. *Science and Technology Studies*, 5(2), 53–62.

Braun, M. 1992. *Picturing Time: The Work of Etienne-Jules Marey (1830–1904)*. Chicago: University of Chicago Press.

Breazeal, C. 2002. *Designing Sociable Robots*. Cambridge, MA: The MIT Press.

Brooks, R.A. 1999. *Cambrian Intelligence: The Early History of the New AI*. London: MIT Press.

Brooks, R.A. 2002. *Robot: The Future of Flesh and Machines*. London: Allen Lane.

Bueno, O. 2006. The Drexler-Smalley debate on nanotechnology: Incommensurability at work?, in *Nanotechnology Challenges: Implications for Philosophy, Ethics, and Society*, edited by J. Schummer and D. Baird. River Edge, NJ: World Scientific, 29–48.

Butler, J. 1993. *Bodies that Matter: On the Discursive Limits of 'Sex.'* New York: Routledge.

Butler, S. 1970. *Erewhon*. Harmondsworth: Penguin.

Bynum, C.W. 1991. *Fragmentation and Redemption: Essays on Gender and the Human Body in Medieval Religion*. Cambridge, MA: Zone Books.

Bynum, C.W. 1995. *The Resurrection of the Body in Western Christianity, 200–1336*. New York: Columbia University Press.

Cameron, D. 2005. Robot, kindly bring me a beer from the fridge. *The Age*, December 3, 21.

Canguilhem, G. 2008. *Knowledge of Life*. New York: Fordham University Press.

Čapek, K. and Čapek, J. 1961. *R.U.R. and the Insect Play*. London: Oxford University Press.

Cartwright, L. 1995. *Screening the Body: Tracing Medicine's Visual Culture*. Minneapolis: University of Minnesota Press.

Cartwright, L. 1997. The Visible Man: The male criminal subject as biomedical norm, in *Processed Lives: Gender and Technology in Everyday Life*, edited by J. Terry and M. Calvert. London: Routledge, 123–37.

Cazort, M. 1996. The theatre of the body, in, *The Ingenious Machine of Nature: Four Centuries of Art and Anatomy*, edited by M. Cazort, M. Kornell, and K.B. Roberts. Ottawa: National Gallery of Canada, 11–42.

Chalmers, D.J. 2003. Absent qualia, fading qualia, dancing qualia, in *Philosophy of Mind: Contemporary Readings*, edited by T. O'Connor and D. Robb. New York: Routledge, 234–54.

Chalmers, D.J. 2010. The Singularity: A philosophical analysis. *Journal of Consciousness Studies*, 17(9–10), 7–65.

Chalmers, D.J. 2011. A computational foundation for the study of cognition. *Journal of Cognitive Science*, 12(4), 323–57.

Channell, D.F. 1991. *The Vital Machine: A Study of Technology and Organic Life.* New York: Oxford University Press.

Clark, A. 2001. *Mindware: An Introduction to the Philosophy of Cognitive Science.* Oxford: Oxford University Press.

Clark, A. 2008. *Supersizing the Mind: Embodiment, Action, and Cognitive Extension.* Oxford: Oxford University Press.

Clynes, M.E. and Kline, N.S. 1960. Cyborgs and space. *Astronautics* (September), 26–7, 74–6.

Conover, M.R. and Miller, D.E. 1981. Elicitation of vocalizations and pecking in ring-billed gull chicks. *Behaviour*, 77(4), 268–86.

Cook, M. and Mineka, S. 1990. Selective associations in the observational conditioning of fear in rhesus monkeys. *Journal of Experimental Psychology: Animal Behavior Processes*, 16(4): 372–89.

Coskun, A. et al. 2012. Great expectations: Can artificial molecular machines deliver on their promise? *Chemical Society Reviews*, 41(1), 19–30.

Crary, J. 1990. *Techniques of the Observer: On Vision and Modernity in the Nineteenth Century.* Cambridge, MA: MIT Press.

Crary, J. 1999. *Suspensions of Perception: Attention, Spectacle, and Modern Culture.* London: MIT Press.

Crevier, D. 1993. *AI: The Tumultuous History of The Search for Artificial Intelligence,* New York: Basic Books.

Crichton, M. 2008. *Prey.* New York: Harper Paperbacks.

Cunningham, A. 1997. *The Anatomical Renaissance: The Resurrection of the Anatomical Projects of the Ancients.* Aldershot, England: Scolar Press.

Dagognet, F. 1992. *Etienne-Jules Marey: A Passion for the Trace.* New York: Zone Books.

Damasio, A.R. 1999. *The Feeling of What Happens: Body and Emotion in the Making of Consciousness.* New York: Harcourt Brace.

Damasio, A.R. 2000. *Descartes' Error: Emotion, Reason and the Human Brain.* New York: Quill.

Daston, L. and Galison, P. 1992. The image of objectivity. *Representations* 40, 81–128.

Dawkins, R. 2006. *The Selfish Gene.* Oxford: Oxford University Press.

de Panafieu, C.W. 1984. Automata: A masculine utopia, in *Nineteen Eighty-Four: Science Between Utopia and Dystopia,* edited by E. Mendelsohn and H. Nowotny. Dordrecht: D. Reidel Publishing Company, 127–45.

de Solla Price, D.J. 1964. Automata and the origins of mechanism and mechanistic philosophy. *Technology and Culture*, 5(1), 9–23.

Deleuze, G. and Guattari, F. 1983. *Anti-Oedipus: Capitalism and Schizophrenia*. New York: Viking.

Deleuze, G. and Guattari, F. 1987. *A Thousand Plateaus: Capitalism and Schizophrenia* B. Massumi, ed. Minneapolis: University of Minnesota Press.

Dennett, D.C. 1981. Where am I?, in *The Mind's I: Fantasies and Reflections on Self and Soul*, edited by D.R. Hofstadter and D.C. Dennett. New York: Basic Books, 217–29.

Dennett, D.C. 1998. *Brainchildren:Essays on Designing Minds*. MIT Press.

Descartes, R. 1972. *Treatise of Man*. Cambridge, MA.: Harvard University Press.

Descartes, R. 1988a. Discourse on the method, in *Descartes: Selected Philosophical Writings*. New York: Cambridge University Press, 20–56.

Descartes, R. 1988b. Meditations on first philosophy, in *Descartes: Selected Philosophical Writings*. New York: Cambridge University Press, 73–122.

Descartes, R. 1988c. Passions of the Soul, in *Descartes: Selected Philosophical Writings*. New York: Cambridge University Press, 218–38.

Drexler, K.E. 1981. Molecular engineering: An approach to the development of general capabilities for molecular manipulation. *Proceedings of the National Academy of Sciences of the United States of America*, 78(9), 5275–8.

Drexler, K.E. 1990. *Engines of Creation*. Garden City, NY: Anchor Books.

Drexler, K.E. Peterson, C. and Pergamit, G. 1991. *Unbounding the Future: The Nanotechnology Revolution*. New York: Morrow.

Dumit, J. 2004. *Picturing Personhood*. Princeton: Princeton University Press.

Egmond, F. 2003. Execution, dissection, pain and infamy: A morphological investigation, in *Bodily Extremities: Preoccupations with the Human Body in Early Modern European Culture*, edited by F. Egmond and R. Zwijnenberg. Burlington, VT: Ashgate, 92–127.

Egmond, F. and Zwijnenberg, R. 2003. Introduction, in *Bodily Extremities: Preoccupations with the Human Body in Early Modern European Culture*, edited by F. Egmond and R. Zwijnenberg. Burlington, VT: Ashgate, 1–9.

Elias, N. 2000. *The Civilising Process: Sociogenetic and Psychogenetic Investigations*. Malden, MA: Blackwell.

Ewen, S. 1988. *All Consuming Images: The Politics of Style in Contemporary Culture*. New York: Basic Books.

Ferrari, G. 1987. Public anatomy lessons and the carnival: The anatomy theatre of Bologna. *Past and Present*, 117(1), 50–106.

Feynman, R.P. 1959. Plenty of Room at the Bottom. [Online]. Available at: http://calteches.library.caltech.edu/1976/1/1960Bottom.pdf [accessed: July 11, 2012].

Foerst, A. 1999. Artificial sociability: From embodied AI toward new understandings of personhood. *Technology in Society*, 21(4), 373–86.

Ford, H. 1922. *My Life and Work*. New York: Doubleday.

Fortunati, L. Katz, J.E. and Riccini, R. 2003. Introduction, in *Mediating the Human Body: Technology, Communication, and Fashion*, edited by L. Fortunati, J.E. Katz, and R. Riccini. Mahwah, New Jersey: Lawrence Erlbaum Associates, 1–11.

Foucault, M. 1975. *Discipline and Punish: The Birth of the Prison*. London: Allen Lane.

Foucault, M. 1977. Nietzsche, genealogy, history, in *Language, Counter-Memory, Practice: Selected Essays and Interviews*, edited by D.F. Bouchard. Oxford: Basil Blackwell, 139–64.

Foucault, M. 1978. *The History of Sexuality, Vol. 1: An Introduction*. New York: Vintage.

Foucault, M. 1985. *The History of Sexuality, Vol. 2: The Use of Pleasure*. New York: Random House.

Foucault, M. 1986. *The History of Sexuality, Vol. 3: The Care of the Self*. Harmondsworth: Penguin.

Foucault, M. 1994. *The Birth of The Clinic: An Archaeology of Medical Perception*. New York: Vintage Books.

Fredrikson, M. Annas, P. and Wik, G. 1997. Parental history, aversive exposure and the development of snake and spider phobia in women. *Behaviour Research and Therapy*, 35(1), 23–8.

Freitas, R.A.J. 2005. Nanotechnology, nanomedicine and nanosurgery. *International Journal of Surgery*, 3(4), 243–6.

Freitas, R.A.J. 2006. Economic impact of the personal nanofactory. *Nanotechnology Perceptions: A Review of Ultraprecision Engineering and Nanotechnology*, 2(2), 1–16.

Freitas, R.A.J. and Merkle, R.C. 2004. *Kinematic Self-replicating Machines*. Georgetown: Landes Bioscience/Eurekah.com.

Gainty, C. 2012. 'Going after the high-brows': Frank Gilbreth and the surgical Subject, 1912–1917. *Representations*, 118(1), 1–27.

Galison, P. 1994. The ontology of the enemy: Norbert Wiener and the cybernetic vision. *Critical Inquiry*, 21(1), 228–66.

Galison, P. 2006. Nanofacture. In C.A. Jones, ed. *Sensorium: Embodied Experience, Technology, and Contemporary Art*. Cambridge, MA: MIT Press, 171–3.

Gallagher, S. 1986. Body image and body schema: A conceptual clarification. *The Journal of Mind and Behavior*, 7(4), 541–4.

Gallagher, S. 1995. Body schema and intentionality, in *The Body and the Self*, edited by J.L. Bermúdez, N. Eilan, and A. Marcel. Cambridge, MA: The MIT Press, 225–44.

Gallagher, S. 2005. *How the Body Shapes the Mind*. Oxford: Clarendon.

Gombrich, E.H. 1948. Icones symbolicae: The visual image in neo-Platonic thought. *Journal of the Warburg and Courtauld Institutes*, 11, 163–92.

Good, I.J. 1966. Speculations concerning the first ultraintelligent machine. *Advances in Computers*, 6, 31–88.

Goodsell, D.S. 2006. Seeing the nanoscale. *Nano Today*, 1(3), 44–9.

Grant, P. 1990. Mobile robotics: Moving away from the intellectual. *Computing and Control Engineering Journal*, 1(4), 167–72.

Greenfield, S. 2012. The Singularity: Commentary on David Chalmers. *Journal of Consciousness Studies*, 19(1–2), 86–95.

Grossman, L. 2011. 2045: The year man becomes immortal. *Time*, February 10.

Grosz, E. 1994. *Volatile Bodies: Toward a Corporeal Feminism*. St. Leonards, NSW: Allen and Unwin.

Gunderson, K. 1964. Descartes, La Mettrie, language, and machines. *Philosophy*, 39(149), 193–222.

Guthold, M. et al. 2000. Controlled manipulation of molecular samples with the nanomanipulator. *IEEE/ASME transactions on mechatronics*, 5(2), 189–98.

Hall, J.S. 1993. *Utility Fog: The Stuff that Dreams are Made of*. [Online]. Available at: http://www.kurzweilai.net/utility-fog-the-stuff-that-dreams-are-made-of [accessed: July 2, 2013].

Hall, J.S., 2001. *What I Want to be When I Grow Up, is a Cloud*. [Online]. Available at: http://www.kurzweilai.net/what-i-want-to-be-when-i-grow-up-is-a-cloud [accessed: July 2, 2013].

Haraway, D.J. 1991. *Simians, Cyborgs, and Women*. New York: Routledge.

Harcourt, G. 1987. Andreas Vesalius and the anatomy of antique sculpture. *Representations* 17, 28–61.

Hayles, N.K. 1998. The condition of virtuality, in *The Digital Dialectic: New Essays on New Media*, edited by P. Lunenfeld. Cambridge, MA: The MIT Press, 68–94.

Hayles, N.K. 1999. *How We Became Posthuman: Virtual Bodies in Cybernetics, Literature, and Informatics*. Chicago: University of Chicago Press.

Hayles, N.K. 2002. Flesh and metal: Reconfiguring the mindbody in virtual environments. *Configurations*, 10(2), 297–320.

Hayworth, K.J. 2012. Electron imaging technology for whole brain neural circuit mapping. *International Journal of Machine Consciousness*, 4(1), 87–108.

Hegel, G.W.F. 1975. *The Philosophy of Fine Art*. New York: Hacker Art Books.

Hegel, G.W.F. 1993. *Introductory Lectures on Aesthetics*. London: Penguin Books.

Hennig, J. 2006. Changes in the design of scanning tunneling microscopic images from 1980 to 1990, in *Nanotechnology Challenges: Implications for Philosophy, Ethics, and Society*, edited by J. Schummer and D. Baird. River Edge, NJ: World Scientific, 143–63.

Hessenbruch, A. 2004. Nanotechnology and the negotiation of novelty, in *Discovering the Nanoscale*, edited by by D. Baird, A. Nordmann and J. Schummer. Amsterdam: IOS Press, 135–44.

Hillier, M. 1988. *Automata and Mechanical Toys: An Illustrated History*. London: Bloomsbury.

Hirschauer, S. 2006. Animated corpses: Communicating with post mortals in an anatomical exhibition. *Body and Society*, 12(4), 25–52.

Hobbes, T. 1996. *Leviathan* 2nd ed. Cambridge: Cambridge University Press.

Hopkins, P.D. 2012. Why uploading will not work, or, the ghosts haunting transhumanism. *International Journal of Machine Consciousness*, 4(1), 229–43

Huyssen, A. 1986. *After the Great Divide: Modernism, Mass Culture, Postmodernism*. Bloomington: Indiana University Press.

Johnson, M. 2007. *The Meaning of the Body: Aesthetics of Human Understanding*. Chicago: University of Chicago Press.

Johnston, J. 2008. *The Allure of Machinic Life: Cybernetics, Artificial Life, and the New AI*. Cambridge, MA: MIT Press.

Jordanova, L.J. 1989. *Sexual Visions*. Madison, WI: University of Wisconsin Press.

Kalil, T. and Wadia, C. 2011. Materials Genome Initiative: A Renaissance of American Manufacturing. [Online]. Available at: http://www.whitehouse.gov/blog/2011/06/24/materials-genome-initiative-renaissance-american-manufacturing [accessed: July 16, 2013].

Kay, L.E. 1997. Cybernetics, information, life: The emergence of scriptural representations of heredity. *Configurations*, 5(1), 23–92.

Kay, L.E. 2000. *Who Wrote the Book of Life?* Stanford: Stanford University Press.

Keller, E.F. 1995. *Refiguring Life: Metaphors of Twentieth-Century Biology*. New York: Columbia University Press.

Keller, E.F. 2000. *The Century of the Gene*. Harvard: Harvard University Press.

Keller, E.F. 2007. Booting up baby, in *Genesis Redux: Essays in the History and Philosophy of Artificial Life*, edited by J. Riskin. Chicago: University of Chicago Press, 334–45.

Kemp, M. 1970. A drawing for the *Fabrica*; and some thoughts upon the Vesalius muscle-men. *Medical History*. 14(3), 277.

Kemp, M. 2006. *Seen/Unseen: Art, Science, and Intuition from Leonardo to the Hubble Telescope*. Oxford: Oxford University Press.

Klestinec, C. 2011. *Theaters of Anatomy*. Baltimore: Johns Hopkins University Press.

Kristeva, J. 1982. *Power of Horror: An Essay on Abjection*. New York: Columbia U.P.

Kurzweil, R. 1999. *The Age of Spiritual Machines: When Computers Exceed Human Intelligence*. New York: Viking.

Kurzweil, R. 2005. *The Singularity is Near: When Humans Transcend Biology*. New York: Viking.

Kurzweil, R. 2006. Reinventing humanity: The future of human-machine intelligence. *The Futurist* (March–April), 39–46.

Kurzweil, R. and Grossman, T. 2004. *Fantastic Voyage: Live Long Enough to Live Forever*. Emmaus, Pa.: Rodale.

Kusuda, Y. 2002. The humanoid robot scene in Japan. *Industrial Robot*, 29(5), 412–19.

Küchler, S. 2008. Technological materiality: Beyond the dualist paradigm. *Theory, Culture and Society*, 25(1), 101–20.

La Mettrie, J.O. de, 1927. *L'homme machine*. Chicago: The Open Court.

Lacan, J. 1977. *Écrits*. London: Tavistock.

Lakoff, G. and Johnson, M. 1980. *Metaphors We Live By*. Chicago: University of Chicago Press.

Landes, J.B. 2007. The anatomy of artificial life: An eighteenth century perspective, in *Genesis Redux: Essays in the History and Philosophy of Artificial Life*, edited by J. Riskin. Chicago: University of Chicago Press, 96–116.

Langton, C.G. 1996. Artificial Life, in *The Philosophy of Artificial Life*, edited by M.A. Boden. New York: Oxford University Press, 39–94.

Le Breton, D. and Walker, R.S. 1988. Dualism and Renaissance: Sources for a modern representation of the body. *Diogenes*, 36(142), 47–69.

Leder, D. 1990. *The Absent Body*. Chicago: University of Chicago Press.

Lee, C.S. 2002. Tin men. *Time*, December 12.

Lenoir, T. 2007. Techno-humanism: Requiem for the cyborg, in *Genesis Redux: Essays in the History and Philosophy of Artificial Life*, edited by J. Riskin. Chicago: University of Chicago Press, 197–220.

Levy, D. 2007. *Love + Sex with Robots: The Evolution of Human–Robot Relations*. New York: HarperCollins.

Lingis, A. 1983. *Excesses: Eros and Culture*. Albany: State University of New York Press.

Lloyd, S. 2002. The computational universe. *Edge*. [Online]. Available at: http://www.edge.org/documents/archive/edge106.html#computational [accessed: July 11, 2012].

Lloyd, S. 2006. *Programming the Universe: A Quantum Computer Scientist Takes on the Cosmos*. New York: Knopf.

Lock, M. and Farquhar, J. 2007. *Beyond the Body Proper: Reading the Anthropology of Material Life*. Durham, NC: Duke University Press.

Locke, J. 1965. *An Essay Concerning Human Understanding*. London: Dent.

Maisano, S. 2007. Infinite gesture: Automata and emotions in Descartes and Shakespeare, in *Genesis Redux: Essays in the History and Philosophy of Artificial Life*, edited by J. Riskin. Chicago: University of Chicago Press, 62–84.

Marra, M. 1999. *Modern Japanese Aesthetics: A Reader*. Honolulu: University of Hawai'i Press.

Marx, K. 1976. *Capital: A Critique of Political Economy*. Harmondsworth: Penguin Books in association with New Left Review.

Maxwell, J.C. 1873. Molecules. *Nature* (September), 437–41.

Mellor, P.A. and Shilling, C. 1997. *Re-Forming the Body: Religion, Community and Modernity*. London: SAGE.

Merleau-Ponty, M. 2002. *Phenomenology of Perception*. London: Routledge.

Metropolis (dir. Fritz Lang, 1927).

Mody, C.C.M. 2006. Small, but determined: Technological determinism in nanoscience, in *Nanotechnology Challenges: Implications for Philosophy, Ethics, and Society,* edited by J. Schummer and D. Baird. River Edge, NJ: World Scientific, 95–130.

Moran, M.E. 2006. The da Vinci robot. *Journal of Endourology,* 20(12), 986–90.

Moran, M.E. 2007a. Jacques Vaucanson: The father of simulation. *Journal of Endourology,* 21(7), 679–83.

Moran, M.E. 2007b. Rossum's Universal Robots: Not the machines. *Journal of Endourology,* 21(12), 1399–402.

Moravec, H.P. 1988. *Mind Children: The Future of Robot and Human Intelligence.* Cambridge, MA: Harvard University Press.

Moravec, H.P. 1999. *Robot: Mere Machine to Transcendent Mind.* New York: Oxford University Press.

Mori, M. 1970. Bukimi no tani [Uncanny Valley]. *Energy,* 7(4), 33–5.

Nagel, T. 1974. What is it like to be a bat? *The Philosophical Review,* 83(4), 435–50.

Nietzsche, F. 2000. On the genealogy of morals, in *Basic Writings of Nietzsche.* New York: Modern Library, 437–599.

Nilsson, N.J. 1984. *Shakey the Robot (Technical note 323),* SRI International. [Online]. Available at: http://handle.dtic.mil/100.2/ADA458918 [accessed: June 5, 2013].

Öhman, A. and Mineka, S. 2003. Fears, phobias, and preparedness: Toward an evolved module of fear and fear learning. *Psychological Review,* 108(3), 483–522.

Ott, J. 2005. Iron Horses: Leland Stanford, Eadweard Muybridge, and the industrialised eye. *Oxford Art Journal,* 28(3), 407–28.

Panksepp, J. 2005. Affective consciousness: Core emotional feelings in animals and humans. *Consciousness and Cognition,* 14(1), 30–80.

Peterson, C. 2010. Freitas awarded first mechanosynthesis patent. [Online]. Available at: http://www.foresight.org/nanodot/?p=3855 [accessed: June 5, 2013].

Phoenix, C. and Drexler, K.E. 2004. Safe exponential manufacturing. *Nanotechnology,* 15(8), 869–72.

Pickering, A. 2004. The science of the unknowable: Stafford Beer's cybernetic informatics, in *The History and Heritage of Scientific and Technological Information Systems,* edited by W.B. Rayward and M.E. Bowden. Medford, NJ: Information Today, Inc., 29–38.

Plato, 1993. *The Last Days of Socrates: Euthypho, Apology, Crito, Phaedo.* London: Penguin.

Porter, R. 2003. *Flesh in the Age of Reason.* London: Allen Lane.

Ptolemy, B. 2009. *Transcendent Man.* Ptolemaic Productions. [Online]. Available at: http://transcendentman.com/ [accessed: June 5, 2013].

Rabinbach, A. 1990. *The Human Motor: Energy, Fatigue, and The Origins of Modernity.* New York: Basic Books.

Ramachandran, V.S. and Hirstein, W. 1999. The science of art: A neurological theory of aesthetic experience. *Journal of Consciousness Studies*, 6(6–7), 15–51.

Ridley, M. 2003. *Nature Via Nurture: Genes, Experience, and What Makes Us Human.* New York: HarperCollins.

Rincon, P. 2004. Nanotech guru turns back on "goo." [Online]. Available at: http://news.bbc.co.uk/2/hi/science/nature/3788673.stm [accessed: July 11, 2012].

Riskin, J. 2003a. Eighteenth-century wetware. *Representations* 83, 97–125.

Riskin, J. 2003b. The defecating duck, or, the ambiguous origins of Artificial Life. *Critical Inquiry*, 29(4), 599–633.

Riskin, J. 2007. Introduction: The Sistine gap, in *Genesis Redux: Essays in the History and Philosophy of Artificial Life*, edited by J. Riskin. Chicago: University of Chicago Press, 1–32.

Rizzolatti, G., Sinigaglia, C. and Anderson, F. 2008. *Mirrors in the Brain: How Our Minds Share Actions, Emotions, and Experience.* Oxford: Oxford University Press.

Robertson, J. 2010. Gendering humanoid robots: Robo-sexism in Japan. *Body and Society*, 16(2), 1.

Robinson, C. 2012. The role of images and art in nanotechnology. *Leonardo*, 45(5), 455–60.

Roco, M.C. 2003. Nanotechnology: Convergence with modern biology and medicine. *Current Opinion in Biotechnology*, 14(3), 337–46.

Sawday, J. 1995. *The Body Emblazoned: Dissection and the Human Body in Renaissance Culture.* London: Routledge.

Sawday, J. 2007. *Engines of the Imagination: Renaissance Culture and the Rise of the Machine.* London: Routledge.

Schaffer, S. 1994. Babbage's intelligence: Calculating engines and the factory system. *Critical Inquiry*, 21(1), 203–27.

Schodt, F.L. 1990. *Inside the Robot Kingdom: Japan, Mechatronics and the Coming Robotopia.* Tokyo: Kodansha International.

Schummer, J. 2003. Aesthetics of chemical products. *HYLE–International Journal for Philosophy of Chemistry*, 9(1), 73–104.

Schummer, J. 2004. Interdisciplinary issues in nanoscale research, in *Discovering the Nanoscale*, edited by by D. Baird, A. Nordmann and J. Schummer. Amsterdam: IOS Press, 9–20.

Schummer, J. 2005. Reading nano: The public interest in nanotechnology as reflected in purchase patterns of books. *Public Understanding of Science*, 14(2), 163–83.

Schummer, J. 2006. Gestalt switch in molecular image perception: The aesthetic origin of molecular nanotechnology in supramolecular chemistry. *Foundations of Chemistry*, 8(1), 53–72.

Schupbach, W. 1982. *The Paradox of Rembrandt's 'Anatomy of Dr. Tulp'.* London: Wellcome Institute for the History of Medicine.

Schwindt, J-M. 2008. Mind as hardware and matter as software. *Journal of Consciousness Studies*, 15, 5–27.

Screech, T. 1996. *The Lens within the Heart: The Western Scientific Gaze and Popular Imagery in Later Edo Japan*. Cambridge: Cambridge University Press.

Segal, D. 2013. This Man Is Not a Cyborg. Yet. *New York Times*. [Online]. Available at: http://www.nytimes.com/2013/06/02/business/dmitry-itskov-and-the-avatar-quest.html [accessed: July 2, 2013].

Seltzer, M. 1992. *Bodies and Machines*. New York: Routledge.

Sheets-Johnstone, M. 1999. *The Primacy of Movement*. Philadelphia: John Benjamins Publishing Company.

Shieber, S.M. 2004. *The Turing Test: Verbal Behavior as the Hallmark of Intelligence*. Cambridge, MA: MIT Press.

Shilling, C. 1993. *The Body and Social Theory*. London: SAGE.

Shimasawa, M. and Hosoyama, H. 2004. *Economic Implications of an Aging Population: The Case of Five Asian Economies*. Tokyo: Economic and Social Research Institute, Cabinet Office.

Shocker (dir. Wes Craven, 1989).

Shusterman, R. 2000. *Performing Live: Aesthetic Alternatives for the Ends of Art*. Ithaca, NY: Cornell University Press.

Slaughter, V. Heron, M. and Sim, S. 2002. Development of preferences for the human body shape in infancy. *Cognition*, 85(3), B71–B81.

Smalley, R.E. 2001. Of chemistry, love and nanobots. *Scientific American*, 285(3), 76.

Smalley, R., Drexler, K.E. and Baum, R., 2003. Point-counterpoint: Nanotechnology. *Chemical & Engineering News*, 81(48), 37–42.

Sofoulis, Z. 2002. Cyberquake: Haraway's manifesto, in *Prefiguring Cyberculture: An Intellectual History*, edited by D. Tofts, A. Jonson and A. Cavallaro. Sydney: Power Publications, 84–103.

Spitzer, V.M. 2002. The Visible Human: Anatomy you can grow with. *The Visible Human Journal of Endosonography*, 1(1).

Stafford, B.M. 1991. *Body Criticism: Imaging the Unseen in Enlightenment Art and Medicine*. Cambridge, MA: MIT Press.

Stallybrass, P. and White, A. 1986. *The Politics and Poetics of Transgression*. London: Methuen.

Suchman, L.A. 2007. *Human–Machine Reconfigurations: Plans and Situated Actions* 2nd ed. Cambridge: Cambridge University Press.

Synnott, A. 1992. Tomb, temple, machine and self: The social construction of the body. *The British Journal of Sociology*, 43(1), 79–110.

Tanaka, F., Cicourel, A. and Movellan, J.R. 2007. Socialization between toddlers and robots at an early childhood education center. *Proceedings of the National Academy of Sciences*, 104(46), 17954–8.

Tarr, S. and Weiss, P.S. 2012. Very small horses: Visualizing motion at the nanoscale. *Leonardo*, 45(5), 439–45.

The Terminator (dir. James Cameron, 1984).

Terminator 2: Judgement Day (dir. James Cameron, 1991).

Terminator Salvation (dir. McG, 2009).

Thacker, E. 2004. *Biomedia*. Minneapolis: University of Minnesota Press.

Thomson, W. 1911. *Mathematical and Physical Papers*. Cambridge: Cambridge University Press.

Tinbergen, N. 1989. *The Study of Instinct*. Oxford: Clarendon Press.

Treichler, P.A, Cartwright, L. and Penley, C., eds. 1998. *The Visible Woman: Imaging Technologies, Gender, and Science*. New York: New York University Press.

Trilling, L. 1972. *Sincerity and Authenticity*. London: Oxford University Press.

Tuomi, I. 2002. The lives and death of Moore's Law. *First Monday*, 7(11), 4.

Tuomi, I. 2003. Kurzweil, Moore, and accelerating change, in *Accelerating Change Conference, Stanford*, 12–14

Turing, A.M. 1950. Computing machinery and intelligence. *Mind*, 59(236), 433–60.

Turkle, S. 1995. *Life on the Screen: Identity in the Age of the Internet*. New York: Simon and Schuster.

van Dijck, J. 2000. Digital cadavers: The visible human project as anatomical theater. *Studies in History and Philosophy of Science Part C*, 31(2), 271–85.

van Dijck, J. 2005. *The Transparent Body: A Cultural Analysis of Medical Imaging*. Seattle: University of Washington Press.

Vartanian, A. 1973. Man-machine from the Greeks to the computer, in *Dictionary of the History of Ideas: Studies of Selected Pivotal Ideas*, edited by P.P. Wiener. New York: Scribner, 131–46.

Villiers de l'Isle-Adam, A. 1982. *Tomorrow's Eve*. Urbana: University of Illinois Press.

Vinge, V. 1993. The coming technological singularity. *Whole Earth Review* 81, 88–95.

Virtuosity (dir. Brett Leonard, 1995).

von Neumann, J. 1966. *Theory of Self-Reproducing Automata*. Urbana: University of Illinois Press.

Waldby, C. 1997. Revenants: The Visible Human Project and the digital uncanny. *Body and Society*, 3(1), 1–16.

Waldby, C. 2000. *The Visible Human Project: Informatic Bodies and Posthuman Medicine*. London: Routledge.

Walter, T. 2004. Body Worlds: Clinical detachment and anatomical awe. *Sociology of Health and Illness*, 26(4), 464–88.

Wiener, N. 1961. *Cybernetics: Or Control and Communication in the Animal and the Machine*. Cambridge, MA: M.I.T. Press.

Wilson, E.A. 2002. Imaginable computers: Affects and intelligence in Alan Turing, in *Prefiguring Cyberculture: An Intellectual History*, edited by D. Tofts, A. Jonson and A. Cavallaro. Sydney: Power Publications, 38–51.

Wilson, E.A. 2010. *Affect and Artificial Intelligence*. Seattle: University of Washington Press.

Wilson, H.R. Loffler, G. and Wilkinson, F. 2002. Synthetic faces, face cubes, and the geometry of face space. *Vision Research*, 42(27), 2909–23.

Wilson, L. 1987. William Harvey's Prelectiones: The performance of the body in the Renaissance Theater of anatomy. *Representations* 17, 62–95.

Wise, M.N. 2007. The gender of automata in Victorian Britain, in *Genesis Redux: Essays in the History and Philosophy of Artificial Life*, edited by J. Riskin. Chicago: University of Chicago Press, 163–95.

Wolfe, J. 2004. *Humanism, Machinery, and Renaissance Literature*. Cambridge: Cambridge University Press.

Wood, G. 2002. *Edison's Eve: A Magical History of the Quest for Mechanical Life*. New York: Alfred A. Knopf.

Yamaguchi, M. 2002. Karakuri: The ludic relationship between man and machine in Tokugawa Japan, in *Japan at Play*, edited by J. Hendry and M. Raveri. New York: Routledge, 72–83.

Ziemke, T. 2007. The embodied self: Theories, hunches and robot models. *Journal of Consciousness Studies*, 14(7), 167–79.

Index